T0227444

Telecommunications Expense Management

How to Audit Your Bills, Reduce Expenses and Negotiate Favorable Rates

by Michael Brosnan, John Messina & Ellen Block

CRC Press
Taylor & Francis Group
Boca Raton London New York

CRC Press is an imprint of the
Taylor & Francis Group, an **informa** business

CRC Press
Taylor & Francis Group
6000 Broken Sound Parkway NW, Suite 300
Boca Raton, FL 33487-2742

First issued in hardback 2017

Copyright © 1999 Michael Brosnan, John Messina and Ellen Block
CRC Press is an imprint of Taylor & Francis Group, an Informa business

No claim to original U.S. Government works

ISBN 13: 978-1-138-41232-3 (hbk)
ISBN 13: 978-1-57820-032-0 (pbk)

This book contains information obtained from authentic and highly regarded sources. Reason-able efforts have been made to publish reliable data and information, but the author and publisher cannot assume responsibility for the validity of all materials or the consequences of their use. The authors and publishers have attempted to trace the copyright holders of all material reproduced in this publication and apologize to copyright holders if permission to publish in this form has not been obtained. If any copyright material has not been acknowledged please write and let us know so we may rectify in any future reprint.

Except as permitted under U.S. Copyright Law, no part of this book may be reprinted, reproduced, transmitted, or utilized in any form by any electronic, mechanical, or other means, now known or hereafter invented, including photocopying, microfilming, and recording, or in any information storage or retrieval system, without written permission from the publishers.

Trademark Notice: Product or corporate names may be trademarks or registered trademarks, and are used only for identification and explanation without intent to infringe.

Visit the Taylor & Francis Web site at
http://www.taylorandfrancis.com

and the CRC Press Web site at
http://www.crcpress.com

Courses, Contact Information & Acknowledgements

Courses

The authors offer seminars and training courses based on the information contained in this book. Sample courses include:

• Telecommunications Bill Auditing Practices
• How to Reduce Your Telecommunications Expenses
• How to Read a Customer Service Record (CSR)
• Contract Negotiation

Contact Information

For more information about the above courses, contact michael.brosnan@infostrategiesgroup.com or brian.leigh@infostrategiesgroup.com.

Acknowledgements

Special thanks to Jennifer McArthur for her assistance in making this book a reality. We can't thank her enough for the diligence and hard work she put in to complete this book. To Michael Messina, Paul Fogarty, Chuck Fraser, Bob Norian, Donna Spoto, Anne Race, Grace Rotondi and Randi Smaldone – thank you for sharing your knowledge and expertise with us as we wrote this book. And to Brian Leigh, thank you for the contributions you made to the entire book. We couldn't have done it without you.

Table of Contents

Introduction

Airfares, car prices and telephone bills are confusing by design. Getting a good deal requires research and skillful negotiating. Telecommunications expenses can be so multifaceted and complex, even the most careful research and skilled negotiating can fall short of achieving the goal of getting a "good deal". Deals change, they're not static. Rates change, services are installed and removed, service providers merge, it's a very dynamic process fraught with risks that make billing errors commonplace. That's why we wrote this book.

Telecommunications Expense Management is a complete guide to managing all your telecommunications billing. We cover a broad spectrum of products and services from the most basic, often called POTS for plain old telephone service, to trunks, interexchange services, network configurations and fraud protection. More importantly, we explain how to save money on all of them.

Large or small, typical telephone bills are constructed using thousands of codes called FIDs and USOCs; leftovers from legacy systems now in transition. These legacy systems reside in mainframes and were developed by the pre-divestiture Bell System or emulated by independent telephone companies and long distance carriers. Once these billing systems were standard but now they are convoluted.

Rather than trying to imagine how distorted telecommunications billing has become, think about the likelihood that you are being overbilled for services. Suppose you were overbilled for years for products and/or services long since disconnected. And suppose you could recover all that overbilling and implement methods to prevent mistakes from happening again. Better yet, imagine going to your management or Board of Directors with a five or six figure check representing the refund you've arranged to obtain. We see this happen often. This book contains real life examples of refunds like this along with information on how these billing errors were discovered, exposed and resolved.

There's a chapter dedicated to optimizing your telecommunications equipment and services that can show you how to save hundreds, if not thousands, of dollars per month. It covers taking inventory of existing telecom equipment and lines, showing you ways to reduce monthly local and long distance billing.

This book is also an excellent reference guide, with information that includes industry resources and shopping tips for communications services. You'll find listings on local and long distance service providers, international carriers, fiber and cable companies, call centers and international callback.

You'll learn how to use the Internet to find suitable service providers and the best deals. Another chapter is devoted to using the Web to help decipher your phone bill and get tariff filings, regulatory rulings and pricing information. You'll learn about industry publications, training classes, seminars, trade shows and conferences.

The telecommunications industry in the United States generates over $200 billion annually. It is technology driven to a point where new products and services define their market and consumer demand tends to be reactive. An industry this big, this sophisticated, this dominating, can overpower you. This book is a survival guide designed to help you cut through the inertia and get to core value. Once there, plan on saving lots of money for your company.

Background

In our first book, we explained the reorganization that followed the 1/1/84 Bell System divestiture. At that time, seven regional bell operating companies (RBOCs) were formed to handle local and intraLATA telephone traffic exclusively (we'll explain LATAs in Chapter One). InterLATA calling was consigned to interexchange carriers (IXCs) like AT&T and MCI. Each discipline had its own mission and could not encroach on the other's territory.

Today, through mergers and acquisitions, the seven RBOCs are now four. IXCs, who themselves have gone through M&As, now offer local and intraLATA service. And the remaining RBOCs, now referred to as ILECs (for incumbent local exchange carriers), face additional competition from CLECs, or competitive local exchange carriers. By the way, ILECs are anxiously awaiting regulatory and judicial approval to go into the interexchange carrier business – in other words, they want to offer long distance services.

Has this confusing world gotten more confusing? You bet.

The major RBOCs today are Bell Atlantic, SBC, BellSouth and US West. Bell Atlantic recently acquired NYNEX and is about to merge with GTE. SBC recently acquired Pacific Telesis and Ameritech. Qwest is buying US West. Collectively, these four RBOCs control over 95% of the local exchange revenues.

The major IXCs are AT&T, MCI/WorldCom and Sprint. Collectively, these three IXCs control over 95% of the interexchange revenues.

Where we once endorsed a monopoly then opted for its breakup, we now see developing an oligopoly or a condition where the telecommunications industry is coalescing into a few significant supercarriers.

In an effort to minimize confusion as you read through this book, we're going to refer to RBOCs and ILECs as local telephone companies. We'll refer to IXCs as long distance carriers. If we slip in a term or two you don't recognize, just flip back to the Glossary where hopefully you'll find a definition.

We're going to do our best to dispel the myths and remove as much confusion as possible. In doing so, we hope to uncover ways for you to save some money.

1

Understanding
Your Local
Telephone Bill

Your Telephone Number

Your ten digit telephone number (for example 916 555-1234) consists of three separate components:

1	2	3
916	555	1234
NPA	NXX	LINE NUMBER

1. *NPA or Area Code* - There are over 200 area codes in the United States, Canada, Bermuda, the Caribbean, and North Western Mexico. Collectively, these countries or regions are part of the North American Numbering Plan. With the ever-increasing demand for facsimile equipment, wireless telephones, pagers and dedicated Internet access, the number of area codes is rapidly expanding. Historically, the second digit of an area code had to be either a 1 or a 0. (i.e. 212 or 914). Due to the demand for area codes the second digit can now be any number between 0 and 9.

2. *NXX* - denotes a specific switch within a local Telephone Company central office. NXXs are numbered as follows: N can be any number from 2 to 9 and X can be any number between 0 through 9. Often times the term NXX is used interchangeably with CO (central office) or exchange.

3. *Line Number* - The last four digits that give a telephone number its unique identity.

Your Local Billing Area

The Historical Perspective: The 1984 breakup of the Bell System separated AT&T from the Bell Operating Companies (RBOC's). The agreement was complex but most importantly defined what services would be provided by AT&T and those services that would be provided by the Bell Operating Companies. Under the terms of the break-up, AT&T would provide long distance service and the Bell Operating Companies would provide local service. The information below details how local and long distance service was defined under the terms of the break-up. It is important to note that GTE and other independent telephone companies were not limited by the terms of the break-up, neither were MCI or Sprint. However, other IXCs and local companies largely voluntarily abided by the terms of the breakup. MCI and Sprint's strategic goals were to keep AT&T regulated while they gained long distance market share. Therefore, neither MCI nor Sprint wanted to do anything that could lead to less regulation of AT&T.

By the mid-1990s the telecommunications market in the US had significantly changed. AT&T had lost significant long distance market share. The Regional Bell Operating Companies desperately wanted to enter the long distance market. The growth of the Internet blurred the differences between local and long distance service. Cable companies wanted to be able to offer phone and Internet service to consumers. In response to this changing environment the Telecommunications Act of 1996 further de-regulated the telecommunications market. AT&T can now provide local and long distance service. The Act also set the conditions under which the Bell Companies could offer long distance service. At the writing of this book, no Bell Company has yet been granted permission to offer long distance service, though this approval is generally considered to be imminent.

LATAs: The break-up of the Bell System had to define the services the Bell Operating Companies could provide. Until they are granted permission to provide long distance services the following remains true: Your local Telephone Company bills you for calls and services within your local calling area and your LATA. A local calling area is your immediate community and may include contiguous neighboring communities.

Local calls are usually billed collectively under a flat rate or, in certain metropolitan areas, billed in message units. A LATA (Local Access and Transport Area) somewhat resembles a congressional district in that it represents a community of common interest. LATAs were created at the time of the Bell System divestiture in 1984 in order to separate revenues that would remain with RBOCs as opposed to those which would remain with AT&T.

The Local Access portion of a LATA provides the connection necessary between your telephone(s) and the local Telephone Company central office. The Transport Area portion of a LATA deals with the connection between that local central office (switching center) and the next switching center in the hierarchy where connections are made to various long distance networks (these connections are called POPs for Points of Presence). The local Telephone Company handles calls and services within a LATA. Calls or services between LATAs are handled by IXCs (see figure 1.1). This is actually the simple part.

Figure 1.1 Local Calling

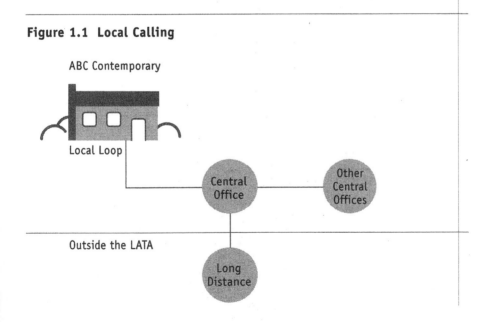

Figure 1.2 LATA's in California

#	City	Area Codes
1	San Francisco	408, 415, 510, 650, 707, 831, 925
2	Chico	530
3	Sacramento	530, 916
4	Fresno	209, 559
5	Los Angeles	213, 310, 323, 424, 562, 615, 626, 714, 760, 805, 818, 909, 949
6	San Diego	619, 760, 858, 935,
7	Bakersfield	661, 805
8	Monterey	831
9	Stockton	209
10	San Luis Obispo	805

LATAs, like congressional districts, are population driven and often default to state and/or metropolitan boundaries. For example, California has eleven LATA's while South Dakota has just one. There are 161 LATAs in the United States. Telephone companies usually will not use the term LATA when describing their local calling areas. In California LATAs are referred to as Service Areas while in New York LATAs are called Regional Calling Areas.

Calls between LATAs in the same state are referred to as intra-state, inter-LATA and are provided by an IXC but regulated by state commissions. Calls between LATAs that cross-state boundaries are also provided by IXCs but are regulated by the FCC.

Figure 1.2 details the LATAs within California.

Your Long Distance Bill

Long distance charges can appear either as part of your local Telephone Company bill or can be billed separately. When they are billed in connection with local telephone charges, they are separately identified and subtotaled.

Long distance charges can be mileage sensitive, time-of-day sensitive, volume sensitive, holiday sensitive, vary between residence and business services, be part of a discount plan and so on. Rates can change with little or no notice. Long Distance charges will be addressed in Chapter 6.

Why Your Local Telephone Bill is Frequently Wrong?

When you order or disconnect a telephone line, there are four basic steps to this process:
1. A local Telephone Company representative enters an order into the Service Order Processing (SOP) system.
2. The order is edited and passed to other departments and downstream provisioning systems like FACS and TIRKS. (see Glossary for definitions).
3. The actual physical work to connect or disconnect a telephone line is completed. Orders are passed to a Billing System, e.g. Customer Record Information System (CRIS).
4. Your bill for services is printed and mailed to you.

A Telephone Company representative first takes your order, then manually assigns USOCs (Universal Service Order Codes) to your request. Each USOC corresponds to a particular Telephone Company product or service. USOC codes also tell the billing system at what price the service should be billed. There are literally thousands of these codes and many are very similar in format but not necessarily similar in price. The more complex the order the greater the chance for errors. Addendum A lists some of the more common USOCs utilized by the major telephone companies.

The representative must also assign an action code to your order so that their billing system knows what to do with the information placed on your order. You will be given an order number such as C2AB1234, D2AB1234, R2AB1234 or N2AB1234. The first letter of your order is your action code.

N – *New order.* The only time an N order is used is when an entirely new account is set up.

C – *Change order.* The most common action code. Indicates a change to existing service. Is used when a service is added (such as an additional telephone line) to existing service or when a service (such as a telephone line) is removed from existing service.

R – *Record Order.* Used when an order only changes telephone company records and does not require physical work to be performed. If you change your billing address an R order is issued.

D – *Disconnect order.* Only used when an entire account is to be disconnected.

While connecting or disconnecting telephone lines may be a four-step process, some telephone companies have over eighty separate systems involved in the ordering, provisioning and billing process. Each one of these systems has their own internal edits or checks. Each system also has its own database of active lines and circuits. These systems often fall out of sync and new service offerings may not be easily entered into one or more of them.

Obsolete codes may be purged from one system but remain active in others. If the order is rejected by one system it is held or frozen until the problem is corrected. Many times, a Telephone Company technician will disconnect a line (perform the physical work) even though the order is held by one of these systems. This is done to avoid work delays. The customer is usually satisfied if the physical work is completed. The order is supposed to be corrected at a later date (to reflect the work already done), but sometimes the volume of new orders becomes so heavy that the problem order is forgotten. In this example your bill never reflects the decrease in price for what you had disconnected.

Special service orders (private lines, DIDs, T1s etc.) also contain many billing errors. Since most circuits connect two locations, it is possible for the billing of a circuit to continue even if one end of it is physically disconnected.

Human errors also contribute to billing problems. The Telephone Company representatives that take your order have a difficult and demanding job. They are required to "overlap" or do two activities at once to increase productivity. For example, while talking to you on the phone, they may be filing documents unrelated to your account. The representative's job is the most demanding and frustrating of all Telephone Company positions. It also has the highest turnover rate.

Another problem for telephone companies involves rate changes. The Telephone Company is constantly applying for increases and/or changes in their rate base. When regulators grant rate changes, they must be applied across all systems and "pro-rated" to reflect proper start dates. Rate changes are often implemented at different times by different systems, each with its own database. The bureaucracy at the Telephone Company often causes one department to be totally unaware of what a system managed by another department is planning to do. The result? Numerous billing errors. This propensity for errors has given birth to a cottage industry of investigators who audit telephone bills and obtain refunds for clients.

The biggest problem is that the systems used by most traditional telephone companies were designed in the 1970s, long before the PC age. These legacy systems are not flexible and were not designed to handle the fast paced world of today.

Understanding Your Local Telephone Bill

A typical local Telephone Company bill summary page follows:

Figure 1.3

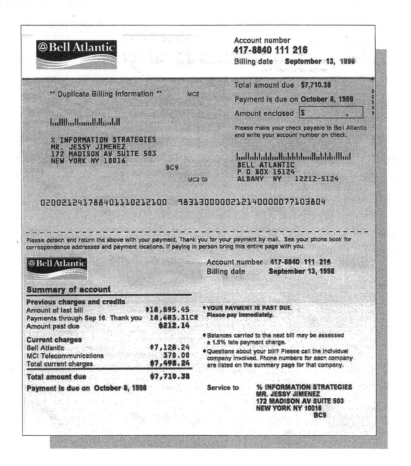

In this example, the customer owes Bell Atlantic $7,710.38. A payment of $18,683.31 was made against the $18.895.45 Bell Atlantic listed as due on last month's bill resulting in a carry over balance of $212.14. Current month charges of $7,128.24 are listed as due for Bell Atlantic provided services. In addition $370.00 worth of charges are identified as due for MCI provided services. Bell Atlantic and other local carriers have agreements in place to bill and collect (for a fee) MCI and other IXC charges under certain circumstances. These charges are known as casual calling charges and are billed at an IXC's highest rates. Though some companies may like the convenience of sending in one check for both local and long distance charges, this approach is not cost justified. As detailed later in the book, casual calling charges should be eliminated by having these calls billed under a calling plan directly by the IXC.

The total current charges due for both Bell Atlantic and MCI charges total $7,498.24. The total amount due is $7,710.38 ($7,498.24 plus the balance left over from the previous month of $212.14).

A further breakdown of Bell Atlantic charges is provided on page 1 of this bill.

Figure 1.4

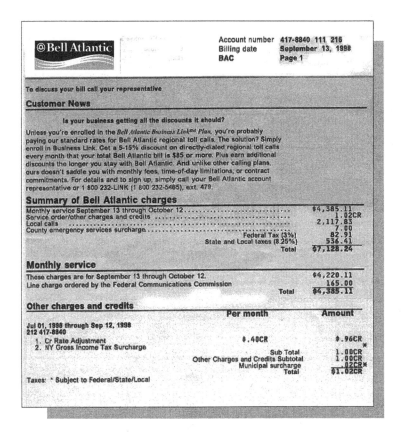

The monthly service September 13 through October 12 charge totals $4,385.11. This charge is for the monthly use of your telephone lines and any other related services such as call waiting.

At the top of this bill the billing date is noted as September 13. This means that on the 13th of every month your bill will be printed. In Bell Atlantic NY the monthly charge for telephone service is billed in advance. The charge is for the period September 13 through October 12. Some companies such as Bell Atlantic – New England bill your monthly service in arrears. In Bell Atlantic – New England the monthly service charge would have been billed from August 13 to September 12.

Local Calls: In all telephone companies local call usage is billed in arrears. Local calls made between August 13 and September 12 are included under this heading. Local calls are those calls made within your LATA. Typically your calls in your local area are not itemized but a summary is provided. Those calls outside your local area but within your LATA are itemized.

A further breakdown is then provided for your monthly service.
a) Monthly charges for September 13 through October 12 total $4220.11.
b) Line charge ordered by the Federal Communications Commission: This charge was approved by the FCC to compensate local telephone companies for loss of subsidies from long distance calling. If you know what the monthly FCC Line Charge is in your state, you can figure out how many lines you have by dividing this number by the total. In Bell Atlantic NY the rate is $8.25 per line (it has recently been changed to $8.13). The total billed under this heading is $165.00. We can determine from this total that this customer has 20 telephone lines ($165.00 divided by $8.25).

Your monthly telephone bill will list your monthly basic service charges but you can not determine if this monthly basic service charge is correct from this limited information. The first thing you need to do is request your Customer Service Record (CSR) from the Telephone Company. The CSR breaks down your basic monthly service charge into all its billable components. When you understand how to read a CSR you will be able to identify hidden and erroneous charges embedded within these charges. Your CSR is obtained by calling the Telephone Company's business office listed on your telephone bill. The Telephone Company will not send out a CSR unless it is requested to do so. It is available free of charge and takes about two weeks to receive. No telecom manager should be without it. The CSR is comprised of USOCs that determine your fixed monthly rental charge.

Customers can call the number listed on their bill to obtain their CSR while professional auditors/consultants have to call special numbers to obtain CSRs on behalf of their clients. Most of the time consultants have to call Telephone Company Consultant Liaison Groups to obtain CSR's. Figure 1.5 lists the telephone and fax numbers for these groups. Consultants need to include a Letter of Agency (LOA) from their client along with the CSR request. An LOA authorizes the consultant to obtain billing information and to negotiate refunds on their customer's behalf.

Figure 1.5 Phone Co. Contact Sheets

Phone Co.	State	Voice #	Fax #
Ameritech	Illinois		630-645-5659
		800-404-4714	414-523-8650
		800-480-8088	800-953-6293
		800-680-6808	
		for all of	Cabs acct. (R)-
		Ameritech	630-645-5660
Ameritech	Indiana		888-570-5958
			630-645-1366
			630-645-0634
		317-556-3546	Cabs acct. (R)-
		800-404-4714	800-635-3706

Phone Co.	State	Voice #	Fax #
Ameritech	Michigan		313-496-9670
			248-975-4318
			Cabs acct. all
		800-482-0647	Letters except (R)-
		800-404-4714* 800-635-3706	
Ameritech	Ohio		313-496-9665
		216-822-3129	888-895-0615
		800-480-2200	800-229-3500
Ameritech	Wisconsin		800-572-6035
			313-496-9670
		800-924-8226	800-924-8226
Bell Atlantic	Delaware	302-761-6151	302-761-6292
		302-576-5228	302-984-1658 lg-b
Bell Atlantic	Maine	800-941-9900	
		800-696-7488	508-884-0995
Bell Atlantic	Maryland		410-393-7270
		410-954-6847	410-393-4662
Bell Atlantic	Massachusetts		508-884-0995
		800-941-9900	508-852-4904
		800-696-7488-lg b	888-603-7772-lg b
Bell Atlantic	New Hampshire		603-669-6536
			800-773-1550
		800-342-9061	800-286-0074-lg b

Phone Co.	State	Voice #	Fax #
Bell Atlantic	New Jersey		973-443-0827
			888-541-9793 Dan
			Cabs acct. (R)-
			973-483-0178
		800-755-1096	609-392-1901
		888-892-5200-lg b	908-754-6848-lg b
Bell Atlantic	Pennsylvania	215-571-6000	215-943-8099
Bell Atlantic	Rhode Island	800-941-9900	508-852-7435
Bell Atlantic	New York		315-732-5680-cabs
	(Upstate)	800-764-9424	607-723-7780
		800-764-9425	607-723-7781
Bell Atlantic	West-Virginia	800-544-5663	800-652-0052
	Virginia		800-474-1010
		800-315-4477	800-652-0052
Bell Atlantic	Washington-DC	800-607-6575-lg b	304-353-5587-lg b
Bell Atlantic	Washington-DC	202-954-6275	703-641-0563
BellSouth	Alabama	800-285-4410	205-972-3945
BellSouth	Georgia	800-452-0255	
		770-452-4320	770-452-5207

Phone Co.	State	Voice #	Fax #
BellSouth	Louisiana		504-592-4950
			225-295-5401
		800-238-5501	800-286-7148
BellSouth	Florida		800-330-5169
		800-753-8172	888-958-4040
		800-296-5545	305-267-8582
BellSouth	Mississippi	800-622-0644	601-961-2938
BellSouth	N-Carolina	800-919-2800	800-214-7985
BellSouth	S-Carolina	800-237-2802	800-679-3943
BellSouth	Tennessee	800-926-2448	800-213-4986
		423-699-2400	423-694-2431
	Kentucky	800-721-8127-cabs	954-958-6330-cabs
BellSouth	All BellSouth	Cabs accts.	800-817-6767
Pacific Bell	California	800-773-3318	626-458-0430
		626-576-3318	626-499-0450
Nevada Bell	Nevada	702-333-4811	702-333-4047
US West	Arizona	800-549-5629	
	New Mexico	800-647-8606	800-868-3351
US West	Colorado		
	Idaho		
	Montana	800-777-9594	800-705-3307

Phone Co.	State	Voice #	Fax #
US West	Iowa		
	Wyoming		
	Minnesota	800-552-1104	800-705-3307
US West	Utah	800-647-8606	801-237-5457
US West	Nebraska		
	N&S Dakota	800-552-1104	800-705-3307
US West	Washington		
	Oregon	800-403-3174	800-252-6418
Southwestern Bell			
	Texas	800-241-0578	800-291-7123
	Oklahoma	214-268-1410	214-464-1611
	Kansas	800-572-9301	800-291-7123
	Missouri	800-241-0578	800-291-7123
Snet	Connecticut		203-925-9456
		800-922-3286	203-925-9739
GTE	Florida	813-623-4000	800-206-0924
	Virginia		800-585-4273
	California	800-344-4831	805-230-3790

USOCs (see Appendix A for a list of common USOCs)

To understand a CSR, you need to understand USOCs. A USOC (pronounced "U-Sock") identifies a particular service or equipment offered by the Telephone Company. USOCs were introduced by AT&T in the late 1970s to provide a platform for commonality among all Bell System companies. AT&T wanted all the Bell Operating Companies to have a common service order processor and billing system. In the 1970s computer technology was very primitive. At that time large mainframe systems dominated the landscape and punch cards were still in use. These systems were unable to recognize "English language" type definitions of telephone services. AT&T had to develop a system where every service was assigned a three or five digit code. These codes became known as USOCs. Bellcore took over this role after the breakup of the Bell System. Bellcore has since been sold and renamed (Telcordia Technologies) and no longer fulfills this role. As you audit bills from different Telephone Companies you will notice many similarities and some differences in the USOC codes utilized by each company.

There are thousands of USOCs in use throughout the former Bell System. Some USOC codes are included in the Telephone Company's tariff (more about tariffs later). Though these companies publish definitions of the USOC codes in use, most of them will not give the definitions directly to customers. Instead of having the convenience of looking up these codes up yourself, you must call the Telephone Company to get each definition.

When you contact the Telephone Company to obtain new service, a representative will take your order and assign the proper USOC. If you order a basic business line with touch tone in New York, a USOC code of 1MB (individual message line - business) or 1FB (individual flat rate line – business) and TTB (touch-tone business) is entered into the Telephone Company's Service Order Processing System. These USOC codes tell the Telephone Company's provisioning and operations departments to install a business line. These USOC codes also tell the billing system to bill the customer business rates as opposed to the less expensive residence rates. (A 1FB USOC code will provide billing at a flat rate as oppose to a 1MB code which will bill for calls on a per message or per minute rate. This option is often offered to residence customers but is less common for business customers). A new residence line would be entered into the Telephone Company's order system as 1MR. Touch-tone on a residence line would be entered as TTR.

The reason why 1MB denotes individual message service business instead of IMB is because the number 1 is always used instead of the letter I in USOC representation. The letter O is also used in place of the numeral 0. USOC codes are either three or five characters and can be alpha, numeric, or a combination of the two. FIDs or field identifiers describe the attributes of a USOC much like an adjective describes a noun. For example, the FID PIC (Primary Interexchange Carrier) will always be somewhere to the right of the USOC 1MB. The FID PIC identifies the long distance carrier on each individual telephone line.

How USOCs Determine Monthly Billing Rates

The local loop is the physical connection between your premises and the Telephone Company's Central Office. This connection is rented from the Telephone Company. The USOC code for the local loop is 1MB. The local loop (often used interchangeably with telephone line) can have features on it such as call waiting or touch-tone. The USOC code for touch-tone is TTB. Each line is also charged a FCC line charge (charge for access to the long distance network). The USOC for the FCC line charge is 9ZR. (The FCC line charge is also called the End User Common Line Charge or EUCL). A word of caution: the FCC line charge rate changes quite frequently. In New York, the current billing for a standard business telephone line is:

1MB	$16.23
TTB	0.00
9ZR	8.13
Total	$24.36

The cost of a basic business telephone line in New York is $24.36. Basic telephone service is referred to as POTS or Plain Old Telephone Service. When you have more than one line at a location, the line under which the bill is rendered is called the Billing Telephone Number or BTN, while each additional telephone line is referred to as an additional or auxiliary line. The additional telephone lines are sometimes called a WTN or Working Telephone Number.

The Customer Service Record (CSR) –Switched and Dedicated Services

Switched Services

Switched Services provide the connection between your company and your local telephone company's Central Office (CO). Calls travel over the local network (called the local loop) to the CO. The CO identifies the call as local (within the LATA) or long distance. Local calls are switched to other local COs as necessary. Long distance calls are switched to your long distance carrier's network. Different services have different USOC codes and therefore different rates. The main types of Switched Service (POTS, Centrex, Trunk, DID & Switched T1) are explained in this chapter.

All of the major local telephone companies (with the exception of Southern New England Telephone) will provide business customers with a copy of their CSR upon request. SNET will take your billing records and type up a summary of the CSR. Without the "raw data" of the CSR, it is difficult to identify certain errors.

Though CSR is a generally recognized term, SNET calls it the Billing Interrogation Record. Some companies will refer to the CSR as the Service Record or Billing Service Record.

The CSR is a copy of how your records appear in the Telephone Company's billing database. It contains all of the

USOCs and FIDs that generate your monthly bill for service. These items are referred to as recurring charges (non-recurring charges are one-time charges for activities such as installation of service). Some of the information on your CSR corresponds to the information taken by the Telephone Company representative during the initial customer contact and were part of the service order entered by them into the SOP system.

This Chapter will utilize CSRs from Bell Atlantic – NY. Bell Atlantic-NY provides a CSR with straight columns. Though CSRs from different companies can look different and information can be in different locations, all contain the same basic information. Once we understand Bell Atlantic-NY CSRs we will provide examples of billing errors from other local and long distance companies later on in this book.

A CSR is divided into different sections: the Header Record Section, the List Section, the Bill Section, and the S&E Section

Figure 2.1 Header Record Section

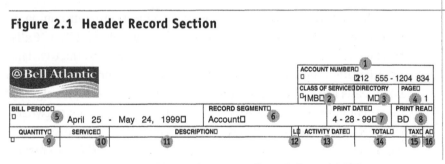

Figures 2.2, 2.2A and 2.2B will detail the List, Bill and S&E sections.

Block	Entry

1   Account Number

The account number, also known as the billing number, is a thirteen digit number that includes a:

three digit Area Code
three digit Central Office prefix
four digit line number, and
three digit customer code

For many customers this number is commonly known as a telephone number

2   Class of Service

The Class of Service block reflects a code that identifies the account service classification. Classifications usually indicate that the account has dial tone access or is a private line with no access to the dial tone network.

3   Directory

This block includes an abbreviated alphabetical identifier for the telephone directory that would ordinarily publish the account name and location.

4   Page

This block includes the page number of the printed paper CSR. All pages are numbered in consecutive ascending order from page 1 through the end of the record.

5   Bill Period

This block reflects the bill date, or billing month. The date includes the month, day, and year.

Header Record Section Legend (Continued)

Block	Entry
6	**Record Segment**
	The CSR has seven sections, commonly known as segments. The segment names are the:
	Account Line & Station (L&S) Key System (KS) Special Service Extra Listings Account Summary S&E Cross Reference
7	**Print Date**
	The print date is the day that the paper record is actually printed.
8	**Print REA**
	The print reason block includes an alphabetical abbreviation that explains why the record was printed. The print reason will always reflect "BD", which means "Bill Date"
9	**Quantity**
	This column reflects the number of items that are listed in the "Service" and "Description" columns.
10	**Service**
	This column reflects the Uniform Service Order Code (USOC) for each service item. Most USOC's will be defined in the "Description" column.

Block	Entry

11   Description

This column may include descriptive information or a definition of the entry in the "Service" column. This column may include, but is not limited to:

primary and secondary (extra) directory listings
Billing Name and Address
Service and Equipment USOC definitions
circuit numbers
glossary of description abbreviations

12   L

The Last activity column reflects the last service order action code that affected the corresponding service item. The action codes are:

Code	Action
E	Enter
I	In
T	To

13   Activity Date

The activity date column reflects the last date on which order activity took place.

14   Total

This column reflects the total monthly charge for the associated "Service" column entry. The total is the product of the number in the "Quantity" column multiplied by the individual rate for the USOC in the "Service" column.

Block	Entry

15   T (shown as "Tax" on your document)

The Tax column contains a numeric tax code that identifies the tax status for an individual service item. The tax codes and corresponding definitions are

Code	Definition
1	Federal and State (local) tax applies
2	No tax applies
3	Federal tax only applies
4	State (local) tax only applies
5	Federal tax applies, State tax applies with local exclusion*
6	State tax applies with local exclusion*

*The term local exclusion means that a portion of the billed charges is exempt from state tax.

16   A

The Activity column will include an asterisk (*) when some action has taken place for the associated item since the last printed CSR was distributed.

List Section (See figure 2.2): This section of the CSR lists important information about your company. It will tell you if your company is listed in the Telephone Directory and how it is listed. It will also tell you the heading under which you are listed (i.e. real estate, restaurants etc.) and how your company is listed with the Directory Assistance operator (411 or 555-1212 operator). If your company name is misspelled here, you could be missing out on quite a lot of business.

Figure 2.2 List, Bill, and S&E Sections

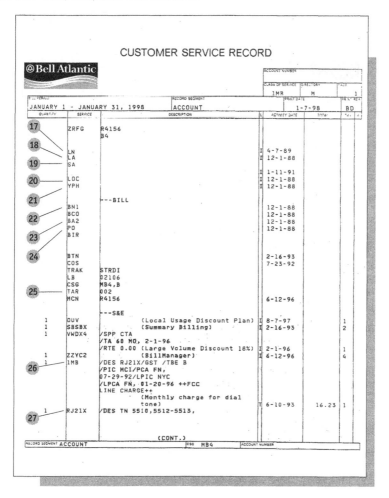

17. *LN - Listed Name.* Denotes how your company is listed in the White Pages and in Directory Assistance. The Service Code NLST instead of LN means that your company has chosen not to be listed in the White Pages Telephone Directory. You get one "free" listing in the White Pages included with your service.

Figure 2.2a S&E Section (continued)

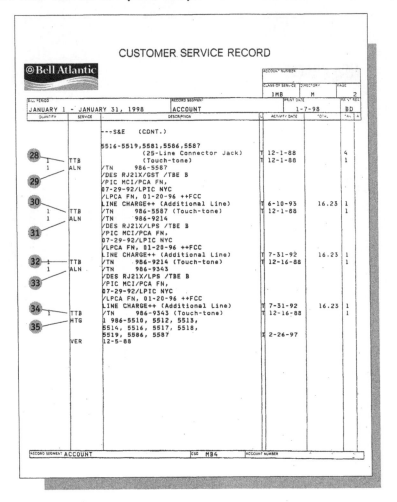

CUSTOMER SERVICE RECORD

18. *LA - Listed Address*. Denotes the address detailed in the White Pages and the address listed with the Directory Assistance operator.

19. *SA - Service Address*. Sometimes the actual physical location (SA) of the telephone service is different from the listed address. The Telephone Company also records the actual location of service.

Figure 2.2b   S&E Section (continued)

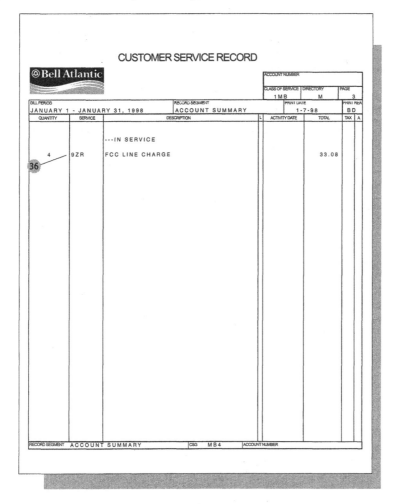

20. *LOC - Location.* Indicates the floor, level and/or the room where the service is physically located.
21. *YPH - Yellow Page Heading.* Utilizes SIC codes to determine the proper Yellow Page Heading for your company. SIC codes are Government issued and classify various types of businesses by function.
22. *BN1-Bill Name.* Lists the party that is responsible for bill payment.

23. *BA - Bill Address.* The address to which your bill is mailed.

24. *PO -Post Office.*

25. *TAR - Indicates the tax area you are in.* This code is utilized by the billing system to identify the correct municipal/city/town/state tax and surcharge rates.

S&E - Service and Equipment Section (See the bottom of Figure 2.2 and all of Figure 2.2A): The name of this section is a throwback to pre-divestiture days when the Telephone Company provided both equipment like telephone sets along with dial tone, touch tone, call waiting etc. (service). Most companies today own or lease their equipment from vendors other than their local telephone companies. In fact, the RBOCs are prohibited from manufacturing telephone equipment under the Divestiture Agreement.

From a pricing/auditing perspective, the S&E section is the most important section because it generates the fixed charges that make up your monthly bill for service. The S&E section lists the USOCs (in the column labeled "SERVICE") and the monthly charge noted for each USOC. If the telephone company representative that took your order entered the wrong USOC, it will show up here. Unless it is discovered and corrected by you, it will most likely remain a perpetual monthly billing error. In later chapters we will detail some of the more common errors that appear on CSRs.

The S&E section generates the monthly bill for service. In this example the service billed is a Basic Business Line. A Basic Business Line or POTS service is the most common form of service provided by your local Telephone Company. This service connects a standard basic telephone set to the Telephone Company CO. Each telephone set or fax machine is connected to its own telephone line and has a unique telephone number assigned to it. POTS service is most common in small companies. The major advantage to POTS service is its simplicity. A POTS line is engineered as "loop start". The electricity that provides power for the line comes from the CO. Dial tone occurs when you lift the phone receiver from the phone releasing the switchhook which completes an electrical circuit (a loop). During an electric blackout your telephone will still work since electricity is provided from the CO. The S&E data shown in this example details the billing for POTS service.

26. *1MB* - We already know that 1MB is the charge for the local loop between the CO and your premises. The monthly charge for this USOC is $16.23. Your 1MB is also the BTN. Since it is always your BTN the description after the USOC 1MB does not identify the telephone number associated with it. You can refer to the heading of the CSR and look in the account number field to obtain the BTN. In this case the 1MB number is 986-5000.

The description field, which contains FIDs (Field Identifiers), describes this USOC in more detail. To the right of the USOC 1MB is a line that reads /PIC MCI/PCA FN. Whenever you see a backward slash (/) the letters following this slash are FIDs. FIDs only appear in the description field of the CSR. The /PIC MCI FID indicates you have told your local telephone company that you want MCI to be your long distance carrier. Each telephone line can have a different long distance carrier. Every long distance carrier is assigned a three digit alpha code and a three digit numeric code (sometimes shown on the CSR with the three-digit code preceded with 10). The CSR may identify the long distance carrier either way after the PIC FID. For example MCI may be identified after the /PIC FID as MCI, 288 or 10288.

The /PCA FN means there is a "freeze" on PIC changes. This customer may have been "slammed" in the past. Slamming is a practice in which a customer's long distance provider is switched without their permission. In short your PIC code is changed without your authorization. The freeze placed on this account means that the local phone company will only take PIC changes from the customer directly. /PCA CN means that the change came from the long distance carrier and /PCA SN means the change came directly from the customer (subscriber).

Checking the CSR is particularly important when you change long distance carriers. The long distance carrier is responsible for contacting the local carrier and having the PIC changed. However, the long distance carrier must identify all of the lines that are to be changed. If a line is not identified it will be billed by the original long distance carrier. There are two ways to determine that each line is PIC'd correctly:

- Order a copy of your CSR and check the PIC on each line.
- Call 1-700-555-4141 to verify your long distance provider. Each verification call must be made from the telephone line being checked.

We will not translate all the FIDs on our sample CSR, as they do not directly impact your bill. As part of your own audit you can call the Telephone Company and obtain the English translation for all USOCs and FIDs that appear on your bill. They are obligated to explain these codes to you. The following are additional explantions of figure 2.2A:

27. *RJ21X* - The RJ21X is a standard Telephone Company jack that allows up to 25 telephone lines to terminate on it. The CSR lists the telephone numbers that terminate on the jack. After listing the full seven-digit telephone (BTN) in the Account Number field, the CSR may only list the last four digits of subsequent telephone numbers. In this example the BTN for this account is 986-5000. The numbers listed after RJ21X are 986-5516, 986-5517, 986-5518, 986-5581, 986-5587. The RJ21X does not generate a monthly recurring charge.

28. *TTB - Touch-tone business.* Most local companies have eliminated the charge for touch-tone.

29. *ALN - Additional or auxiliary line.* The ALN USOC lists the charge for additional POTS lines. /TN stands for telephone number (in this example 986-5587). Sometimes only 5587 will be listed after /TN. Each auxiliary line will also have PIC information listed.

30. *TTB - Touch-tone business.* Touch-tone is provided on line 986-5587.

31. *ALN* - The next telephone number billed to this account is 986-9214.

32. TTB - Touch-tone business. Touch-tone is provided on line 986-9214.

33. *ALN* - The next telephone number billed to this account is 986-9343.

34. *TTB - Touch-tone business.* Touch-tone is provided on line 986-9343.

    Each ALN repeats the same basic information. With POTS service once you can read a CSR with four lines you can read a CSR with 100 lines.

35. *HTG - Hunting or "rollover".* Surprisingly many RBOCs do not charge for hunting, while others charge very dearly for this service. Bell Atlantic – NY does not charge for hunting. Hunting gives you the ability to have a call automatically transferred to another line after a certain number of rings without an answer or if a particular number is busy.

    The proper hunting sequence is critical to a business's success. Yet it is often overlooked or misunderstood. Hunting is often set up when service is initiated but is rarely changed as the business changes. One of our clients told us that she had been receiving complaints from her customers that they would sometimes get the mail room when dialing the main customer service number. She could not understand how this would happen sporadically. Upon examination of her CSR it was discovered that the telephone number for the mailroom was listed at the end of her customer service hunt group. When all lines were simultaneously busy, overflow calls were being "rolled-over" to the mailroom. You can imagine how much business this cost and how easy it is to correct when you can read a CSR.

    This CSR lists the hunt sequence as 986-5510, then 986-5512, 986-5513, 986-5514, 986-5516, 986-5517, 986-5518, 986-5519, 986-5586 and 986-5587. The problem with this hunting set-up is that it will not work. All of the numbers listed except for 986-5587 are no longer on the CSR.

36. *9ZR* - We have previously identified the USOC 9ZR as being the USOC for the FCC line charge. Each telephone line that you have is billed a 9ZR. In this example, the quantity column lists 4, so we know there must be four lines. If there are not four lines billed, then we have a billing error. In this case the four lines are 986-5000 (1MB) and 986-5587, 986-9214 and 986-9343. The total charge listed is $33.08 but since there are four lines the charge per 9ZR is $8.27.

Trunk Verses POTS CSR Billing

Some telephone companies charge more for what is known as a trunk line (Bell Atlantic – NY does not). A trunk line connects two switches (the telephone company's switch in the CO to your PBX switch) and is designed to work with a PBX. A PBX is a computer with the intelligence to route incoming and outgoing calls based on routing tables programmed into its memory.

A trunk is installed be the telephone company as ground start while POTS lines are usually loop start. Ground start is where one side of the trunk line is momentarily grounded (usually to a cold water pipe). Your PBX requires a source of electricity in order to operate. In the event of a blackout or power shortage you need a back-up power supply in order to keep your phones operational.

A PBX allows your company the ability to transfer calls between stations, four-digit dialing, voice mail and other advanced features. Figure 2.3 depicts a standard PBX configuration. With trunk service each employee does not have their own dedicated phone lines but must dial a number (usually 9) to obtain an unused trunk line. Trunks allow access to the local network to be shared as needed by a company's employees. For example, a company with 2000 people at a particular location may only need 200 trunk lines instead of 2000 separate POTS lines. The reduced number of trunks result in savings that offset the expense of a PBX.

How a PBX works: One of your employees picks up their handset and dials the person they wish to call. The PBX analyzes the call and switches the call based on three criteria:

- If the dialed number has only four digits, the PBX routes the call to another extension connected to the PBX. This is an internal call.

- If the dialed number has seven digits, the PBX forwards the call over the local loop to the local Telephone Company.

- If the number is 1 or 0 plus ten digits, the call is routed to the local Telephone Company for delivery to your long distance carrier (as noted by your PIC selection on that account).

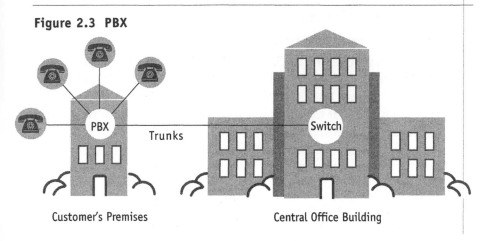

Figure 2.3 PBX

PBX

Trunks

Switch

Customer's Premises

Central Office Building

A trunk can be configured as one way in (allows people to call in but blocks outgoing calls), one way out (blocks incoming calls but allows outgoing calls) or as a combination trunk (allows both incoming or outgoing calls). Each one of these configurations has its own USOC. Bell Atlantic – NY commonly utilizes the following USOCs:

TCG   Initial combo trunk

TXG   Each additional combo trunk

TCM   Initial out only trunk

TXM   Each additional out only trunk

Sample CSR Billing: Trunk Service

The Header, List and Bill Sections are standard across different Switched Service offerings. The S&E Sections, however, are different. Figures 2.4 & 2.4A detail the S&E for trunk service in Bell Atlantic – NY.

The following explantions refer to figures 2.4a & 2.4b:

1. TXG – Trunks will be identified in telephone number format. Often times you will see the followoing format listed here; 96,TKNA,XXX,337,1018. The TKNA identifies the service as trunk service. The XXX will be the area code and 337-1018 is the telephone number of the trunk. Most of the FID information stays the same as POTS. The /PIC is identified as MCI. The CSR describes the service as "Combined Trunk". The cost for the trunk is $16.23.

2. TJB – Touch-tone on trunk service.

3. TXG – The next trunk line is identified as 337-3108.

4. 9ZR - Every trunk line has a FCC charge billed to it. In this example we only listed two of the four trunks billed on this CSR (hence the "4" 9ZR charges). If the CSR listed only two trunks and had four 9ZR charges billed to it then this would be an example of overbilling by the local telephone company.

Figure 2.4a

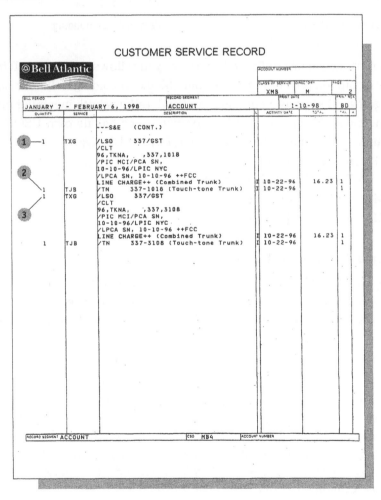

Figure 2.4b

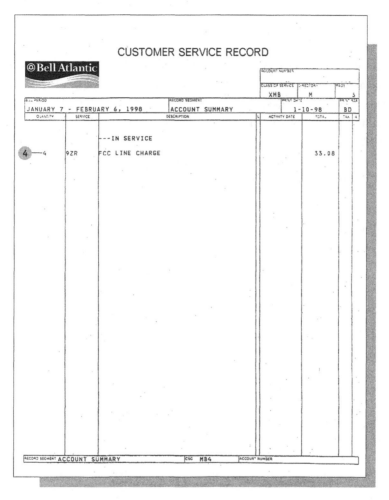

CUSTOMER SERVICE RECORD

Bell Atlantic

QUANTITY	SERVICE	DESCRIPTION	L	ACTIVITY DATE	TOTAL	TAX	A
		---IN SERVICE					
4 —4	9ZR	FCC LINE CHARGE			33.08		

RECORD SEGMENT ACCOUNT SUMMARY — CSG MB4 — ACCOUNT NUMBER

Sample CSR Billing: Centrex Service

Centrex provides an alternate to trunk service. Your local telephone company's CO is configured to allow PBX like services (four-digit dialing between employees, voice mail etc.). It is an alternate for those companies that do not want to incur the initial expense of a PBX and the maintenance costs associated with a PBX. Each employee must still have his or her own telephone number and telephone line.

Digital Centrex has three billable components:
Centrex Station Line – Based on number of lines and contract period.
CEAC Charge - $1.61
FCC Line Charge - $8.13

Centrex can be either analog or digital. Centrex is often provided under a contract and rates vary according to the length of the contract and the number of lines. For Centrex accounts over 100 lines the pricing is usually provided on an individual case basis (ICB). This allows the local Telephone Company to come up with a price that applies to a particular client as long as it is not below the local telephone company's cost. The following is a analog Centrex CSR:

Figure 2.5a

CUSTOMER SERVICE RECORD

Bell Atlantic

ACCOUNT NUMBER	CLASS OF SERVICE	DIRECTORY	PAGE
	ETKUR	M	1

BILL PERIOD	RECORD SEGMENT	PRINT DATE	PRINT REA
MAY 13 - JUNE 12, 1999	ACCOUNT	5-17-99	BD

QUANTITY	SERVICE	DESCRIPTION	L	ACTIVITY DATE	TOTAL	TAX	A
	PCL	LOCL					
	CENT	TAM					
	LN						
			*	3-27-85			
	LA						
			*	3-27-85			
	SA						
			I	7-12-91			
	LOC	FLR 10	I	7-12-91			
	YPH			1-1-76			
		---BILL					
	BN1			**-**-**			
	BN2			**-**-**			
	BA1			**-**-**			
	PO			**-**-**			
	ADV						
	CCH	5					
	LB	03036					
	CSG	T09,B					
	TAR	002					
	MCN	MNM011182C		6-26-98			
		---RMKS					
	RMKR	1-ASSUME DIAL 9, NO ADL FEA	I	3-26-92			
		---S&E					
1	ETKUR		T	7-25-94			
1	BDT	(Toll Free Service)	I	5-30-96	5.00	4	
1	BSX	(Bell Atlantic Calling					
		Card)		4-9-85		2	
1	BSX	(Bell Atlantic Calling					
		Card)	I	8-6-85		2	
1 — 1	NYZAA	/SPP 2P9					
		/TA 120 MO, 07-25-97					
		/CN 01/ZLSZ 00039					
		/RTE 626.34 (Contract Package)	I	11-7-97	626.34	4	
2 — 1	ALN	/TN 212 242-0862					
		/DES EXSTNG RJ21X					
		(CONT.)					

RECORD SEGMENT	CSG	ACCOUNT NUMBER
ACCOUNT	T09	

Figure 2.5b

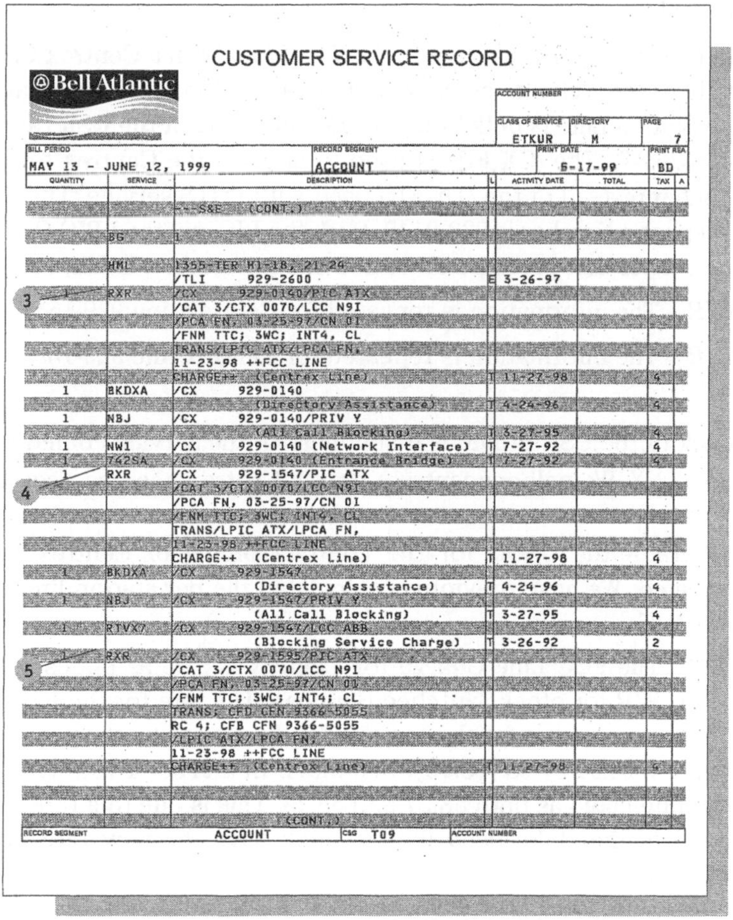

CUSTOMER SERVICE RECORD

Bell Atlantic

ACCOUNT NUMBER

CLASS OF SERVICE: ETKUR | DIRECTORY: M | PAGE: 7

BILL PERIOD: MAY 13 - JUNE 12, 1999 | RECORD SEGMENT: ACCOUNT | PRINT DATE: 6-17-99 | PRINT REA: BD

QUANTITY	SERVICE	DESCRIPTION	L	ACTIVITY DATE	TOTAL	TAX	A
		---S&E (CONT.)					
	BG	1					
	HML	1355-TER H1-18, 21-24					
		/TLI 929-2600	E	3-26-97			
1	RXR	/CX 929-0140/PIC ATX					
		/CAT 3/CTX 0070/LCC N9I					
		/PCA FN, 03-25-97/CN 01					
		/FNM TTC; 3WC; INT4, CL					
		TRANS/LPIC ATX/LPCA FN,					
		11-23-98 ++FCC LINE					
		CHARGE++ (Centrex Line)	T	11-27-98		4	
1	BKDXA	/CX 929-0140					
		(Directory Assistance)	T	4-24-96		4	
1	NBJ	/CX 929-0140/PRIV Y					
		(All Call Blocking)	T	3-27-95		4	
1	NW1	/CX 929-0140 (Network Interface)	T	7-27-92		4	
1	742SA	/CX 929-0140 (Entrance Bridge)	T	7-27-92		4	
1	RXR	/CX 929-1547/PIC ATX					
		/CAT 3/CTX 0070/LCC N9I					
		/PCA FN, 03-25-97/CN 01					
		/FNM TTC; 3WC; INT4, CL					
		TRANS/LPIC ATX/LPCA FN,					
		11-23-98 ++FCC LINE					
		CHARGE++ (Centrex Line)	T	11-27-98		4	
1	BKDXA	/CX 929-1547					
		(Directory Assistance)	T	4-24-96		4	
1	NBJ	/CX 929-1547/PRIV Y					
		(All Call Blocking)	T	3-27-95		4	
1	RTVX7	/CX 929-1547/LCC ABB					
		(Blocking Service Charge)	T	3-26-92		2	
1	RXR	/CX 929-1595/PIC ATX					
		/CAT 3/CTX 0070/LCC N91					
		/PCA FN, 03-25-97/CN 01					
		/FNM TTC; 3WC; INT4; CL					
		TRANS; CFD CFN 9366-5055					
		RC 4; CFB CFN 9366-5055					
		/LPIC ATX/LPCA FN,					
		11-23-98 ++FCC LINE					
		CHARGE++ (Centrex Line)	T	11-27-98		4	
		(CONT.)					

RECORD SEGMENT: ACCOUNT | CSG: T09 | ACCOUNT NUMBER

The following explanations refer to figures 2.5a - 2.5c.

1. *NYZAA* – This is a special USOC used by Bell Atlantic – NY to indicate that a contract is in place. The /SPP FID indicates a Special Pricing Plan. The /TA FID indicates a Term Agreement which is the length of the contract. In this example, the contract was signed on 7-25-97 for 120 months and therefore expires on 7-25-07. The /ZLSZ FID provides the number of Centrex lines. The Centrex Line Charge can be determined by dividing the total cost of $626.34 by 39 lines. The per line charge is $16.06 and must be added to CEAC and FCC Line Charge at the back of the CSR to determine the full cost per line.

2. *ALN – Additional Line.* POTS service can be mixed with Centrex. We will skip over this for now and concentrate on the Centrex portion of the CSR.

3. *RXR – USOC for the Centrex Station Line.* With a Centrex CSR you will not see a charge on each individual Station Line. The charge for the Station Line is bulk billed at the beginning of the CSR (at a total of $626.34). The /CX FID is used to indicate the Centrex telephone number instead of the /TN FID used with POTS service. The Station Telephone Number is 929-0140. Both the PIC and the LPIC (indicates the carrier that provides extended calling within a LATA) are AT&T.

4. *RXR* – The information is repeated for Centrex line 929-1547.

5. *RXR* – The information is repeated for Centrex line 929-1595.

The back of the CSR provides a summary of the services billed by Bell Atlantic – NY. We will concentrate on the Centrex charges. (see Figure 2.5c)

6. *NYZAA* – the USOC for the contract billing of Centrex lines is repeated. We have previously calculated the per station line charge at $16.06.

7. *9ZC* - The FCC Line Charge for Centrex is 9ZC as opposed to the USOC 9ZR for other Switched Services. The rate for the USOC 9ZC is the same as that for the USOC 9ZR. To determine the per line rate divide $308.94 by 38 ($8.13). Reminder: In this book you will see the 9ZR rate listed as both $8.27 and $8.13. This is due to a rate change. Some of the CSRs used in this book have the new rate of $8.13 reflected.

8. 38 CEAC charges at a total of $1.61 are listed.

The total cost per Centrex line is calculated as follows:
NYZAA contract USOC
(per Centrex station line) $16.06
CEAC $ 1.61
9ZC $ 8.13
Total $25.58

Figure 2.5c

CUSTOMER SERVICE RECORD						

@ Bell Atlantic

			ACCOUNT NUMBER			
			CLASS OF SERVICE: ETKUR	DIRECTORY: M		PAGE: 23

BILL PERIOD	RECORD SEGMENT			PRINT DATE		PRINT AREA
MAY 13 - JUNE 12, 1999	ACCOUNT SUMMARY			5-17-99		BD

QUANTITY	SERVICE	DESCRIPTION	L	ACTIVITY DATE	TOTAL	TAX A
		---IN SERVICE				
4	ALN	@ 16.23			64.92	
1	BGT	@ 5.00			5.00	
1	NYZAA	@ 626.34			626.34	
1	VMH2A	@ 12.46			12.46	
1	W1E	@ 10.00			10.00	
	BG 1					
2	VSFAZ	@ 8.50			17.00	
38	9ZC	FCC LINE CHARGE			308.94	
4	9ZR	FCC LINE CHARGE			32.52	
		MUNICIPAL SURCHARGE			25.43	
		N.Y. STATE/MTA SURCHARGE			40.64	
		NY FCC SURCHARGE			13.61	
		---IN SERVICE				
		CENTREX LINE SUMMARY				
0		PORT PRIMARY LINES				
38		PRIMARY LINES				
38		CIC @ 0.00 POST 7-28-83				
38		CEAC @ 1.61			61.18	
0		PORT CEAC @ 0.00				
		CENTREX LINE SUMMARY				
	BG 1					
37	RXR					
1	RX3					

RECORD SEGMENT: ACCOUNT SUMMARY	CSG T09	ACCOUNT NUMBER

The number of CEAC charges and 9ZC charges must match or you have a billing problem. The number of Centrex Station Lines under contract should also match. In some instances a company may sign a contract for a certain number of lines over a certain period of time and later find it needs to downsize. In that case they must pay for all of the NYZAA contract lines but can eliminate the CEAC and 9ZC charges on the unused lines. In this case the lines under contract equal 39 but the number of CEAC and 9ZC charges equal 38. This discrepancy needs to be investigated. Real cases of Centrex billing problems and their resolutions are covered in Chapter 4.

Sample CSR Billing: DID Trunks

Direct Inward Dialing (DID) service provides an outside caller with direct access to a specific individual telephone extension connected to a PBX without the assistance of an attendant. Direct Inward Dialing trunks allow many different telephone numbers to share a trunk (though only one telephone number can use the trunk at any one time). Customers are assigned Direct Inward Dialing phone numbers in blocks of 20 or 100, i.e. 555-1000 to 555-1020.

The components of a DID trunk are similar throughout most local telephone companies. The monthly rates shown below are DID rates and USOCs for Bell Atlantic – NY.

DID numbers:

	USOC	RATE
DID Station Numbers	ND4	$ 3.64 per group of 20 numbers
DID Station Numbers	NDZ	$18.23 per group of 100 numbers

DID trunks:

	USOC	RATE
DID Trunk Charge	TB2	56.17
DID Loop (2 wire)	D1F2X	19.08
FCC Line Charge	9ZR	8.27
Total		$83.52

The total cost of a DID trunk in Bell Atlantic-NY is $83.52. You are charged a premium for the ability of a single DID trunk to recognize many different telephone numbers.

The following is a CSR that itemizes billing for twelve DID trunks and 200 DID numbers:

Figure 2.6a

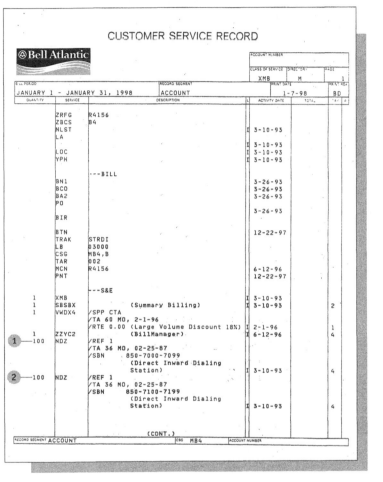

The following explanations refer to figures 2.6a - 2.6d.

1. *NDZ* – USOC for 100 DID numbers, 850-7000 through 850-7099. The cost for this group of numbers is listed in the back of the CSR.

2. *NDZ* – USOC for 100 DID numbers, in this case 850-7100 through 850-7199. The cost for this group of numbers is listed in the back of the CSR.

Figure 2.6b

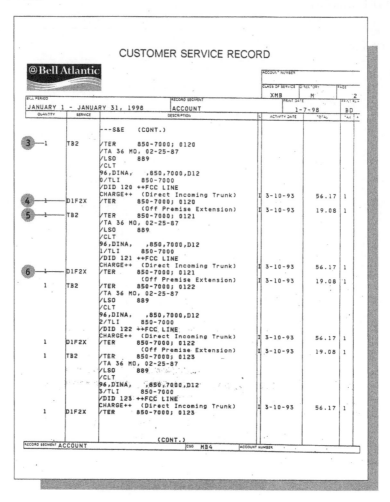

CUSTOMER SERVICE RECORD

Bell Atlantic

					ACCOUNT NUMBER			

CLASS OF SERVICE DIRECTORY XMB M 2

BILL PERIOD: JANUARY 1 – JANUARY 31, 1998   RECORD SEGMENT: ACCOUNT   PRINT DATE 1-7-98   BD

QUANTITY	SERVICE	DESCRIPTION	L	ACTIVITY DATE	TOTAL	TAX
		---S&E (CONT.)				
③—1	TB2	/TER 850-7000; 0120				
		/TA 36 MO, 02-25-87				
		/LSO 889				
		/CLT				
		96,DINA, ,850,7000,D12				
		0/TLI 850-7000				
		/DID 120 ++FCC LINE				
		CHARGE++ (Direct Incoming Trunk)	I	3-10-93	56.17	1
④—1	D1F2X	/TER 850-7000; 0120				
		(Off Premise Extension)	I	3-10-93	19.08	1
⑤—1	TB2	/TER 850-7000; 0121				
		/TA 36 MO, 02-25-87				
		/LSO 889				
		/CLT				
		96,DINA, ,850,7000,D12				
		1/TLI 850-7000				
		/DID 121 ++FCC LINE				
		CHARGE++ (Direct Incoming Trunk)	I	3-10-93	56.17	1
⑥—1	D1F2X	/TER 850-7000; 0121				
		(Off Premise Extension)	I	3-10-93	19.08	1
1	TB2	/TER 850-7000; 0122				
		/TA 36 MO, 02-25-87				
		/LSO 889				
		/CLT				
		96,DINA, ,850,7000,D12				
		2/TLI 850-7000				
		/DID 122 ++FCC LINE				
		CHARGE++ (Direct Incoming Trunk)	I	3-10-93	56.17	1
1	D1F2X	/TER 850-7000; 0122				
		(Off Premise Extension)	I	3-10-93	19.08	1
1	TB2	/TER 850-7000; 0123				
		/TA 36 MO, 02-25-87				
		/LSO 889				
		/CLT				
		96,DINA, ,850,7000,D12				
		3/TLI 850-7000				
		/DID 123 ++FCC LINE				
		CHARGE++ (Direct Incoming Trunk)	I	3-10-93	56.17	1
1	D1F2X	/TER 850-7000; 0123				
		(CONT.)				

RECORD SEGMENT ACCOUNT   CSO MB4   ACCOUNT NUMBER

3. *TB2* – Direct Incoming Trunk Charge. DID trunks are usually identified by the same ficticious trunk number but with a unique suffix such as 001, 002 etc. In this example the DID trunk is identified as /TER 850-7000; 120. We can only assume that at one time this customer had at least 120 DID trunks and that they downsized. Normal convention would have the first DID trunk listed as 850-7000; 001. The charge for the TB2 is $56.17 a month.

Figure 2.6c

```
                    CUSTOMER SERVICE RECORD

    @Bell Atlantic                          ACCOUNT NUMBER

                                     CLASS OF SERVICE  DIRECTORY     PAGE
                                          X M B            M            6
    BILL PERIOD              RECORD SEGMENT          PRINT DATE      SIGNATURE
    JANUARY 1 - JANUARY 31, 1998     ACCOUNT              1-7-98       BD
     QUANTITY    SERVICE          DESCRIPTION      L   ACTIVITY DATE   TOTAL   TAX

                        ---S&E    (CONT.)

        1     D1F2X    /TER      850-7000; 0127
                                 (Off Premise Extension)  I  3-10-93   19.08  1
        1     TB2      /TER      850-7000; 0128
                       /TA 36 MO, 02-25-87
                       /LSO      889
                       /CLT
                       96,DINA,    ,850,7000,D12
                       8/TLI      850-7000
                       /DID 128 ++FCC LINE
                       CHARGE++  (Direct Incoming Trunk)  I  3-10-93   56.17  1
        1     D1F2X    /TER      850-7000; 0128
                                 (Off Premise Extension)  I  3-10-93   19.08  1
        1     TB2      /TER      850-7000; 0129
                       /TA 36 MO, 02-25-87
                       /LSO      889
                       /CLT
                       96,DINA,    ,850,7000,D12
                       9/TLI      850-7000
                       /DID 129 ++FCC LINE
                       CHARGE++  (Direct Incoming Trunk)  I  3-10-93   56.17  1
        1     D1F2X    /TER      850-7000; 0129
                                 (Off Premise Extension)  I  3-10-93   19.08  1
        1     TB2      /TER      850-7000; 0130
                       /TA 36 MO, 02-25-87
                       /LSO      889
                       /CLT
                       96,DINA,    ,850,7000,D13
                       0/TLI      850-7000
                       /DID 130 ++FCC LINE
                       CHARGE++  (Direct Incoming Trunk)  I  3-10-93   56.17  1
        1     D1F2X    /TER      850-7000; 0130
                                 (Off Premise Extension)  I  3-10-93   19.08  1
        1     TB2      /TER      850-7000; 0131
                       /TA 36 MO, 02-25-87
                       /LSO      889
                       /CLT
                       96,DINA,    ,850,7000,D13
                       1/TLI      850-7000
                       /DID 131 ++FCC LINE


                            (CONT.)
    RECORD SEGMENT ACCOUNT            CSG  MB4    ACCOUNT NUMBER
```

4. *D1F2X* - Each DID has one loop charge associated with it. The charge associated with DID Trunk 850-7000; 120 is $19.08 a month.

5. *TB2* – The Direct Incoming Trunk Charge is repeated for each DID trunk. This DID trunk is listed as 850-7000; 121.

6. *D1F2X* - The charge associated with trunk 850-7000; 121 is $19.08 a month.

The TB2 and D1F2X charges are repeated for trunks 850-7000; 122, 850-7000; 123, 850-7000; 124, 850-7000; 125, 850-7000; 126, 850-7000; 127, 850-7000; 128, 850-7000; 129, 850-7000; 130 and 850-7000; 131. These twelve trunks are listed in Figures 2.6b through 2.6d.

Figure 2.6d

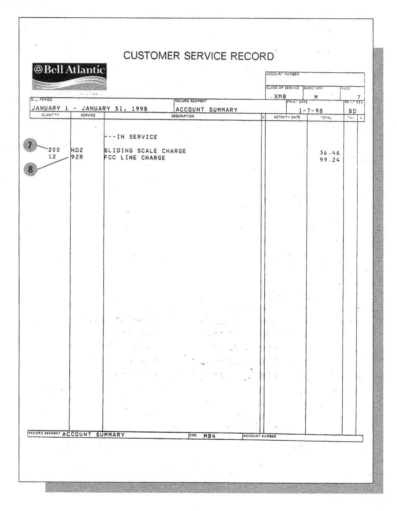

7. 200 NDZ charges are listed at the back of the CSR. The charge for each group of 100 DID numbers is $18.23, so the total charge listed is $36.46.

8. Each DID trunk is billed an FCC line charge. The FCC line charge is billed at the rate of $8.27 per line and the total billed against this USOC is $99.24 (12 x $8.27).

DID USOCs across different telephone companies are very similar. For example, DID USOCs utilized by Pacific Bell include:

ND8   Each group of 100 telephone numbers under first 200
NDA   Each group of 100 telephone numbers above the first 200
TMN   DID trunk charge-assured, measured rate
TCT   Circuit termination charge per DID trunk
9ZR   FCC line charge

The TMN USOC is equivalent to Bell Atlantic-NY's TB2 USOC, while the ND8 USOC is equivalent to Bell Atlantic-NY's NDZ USOC. The TCT USOC is equivalent to Bell Atlantic-NY's D1F2X USOC (2-wire loop charge).

The following USOCs and rates utilized by Bell Atlantic-NJ further illustrate the similarities in DID billing among former Bell System companies:

USOC	Description	Rate
NDS	DID 1ST 20 Line Numbers	$20.00
TDD	/TN 609 555-9999A	$8.00
NDT	/TN 609 555-9999A	$38.97
	/DID Trunk Charge	
9ZR	/FCC Line Charge	$3.82

On a New Jersey Bell CSR, a DID trunk is identified with a telephone number plus an alpha suffix (555-9999A then 555-9999B etc.) as opposed to the numeric suffix assigned on a Bell Atlantic-NY CSR (i.e. 555-9999 001 then 555-9999 002 etc.).

Sample CSR Billing: Switched T1

The advent of digital PBXs brought about a demand for Switched T1 service. Digital PBXs allow a direct T1 connection into the PBX. Each T1 can be separated into twenty-four separate channels. Each channel can carry either one-way, two-way or DID trunks. Trunks on a T1 are said to be "riding" that T1.

Every local telephone company has sought to uniquely brand its Switched T1 with proprietary names such as ADTS, MEGACOM and FLEXPATH. The difference in the service has more to do with how it is billed by each company than in the way the service is provided. In some companies you pay a certain rate for each of the twenty-four trunks (channels) on the T1 as well as a charge for the T1 circuit itself. In others you only pay for the T1 circuit. The individual trunks are included in the T1 price.

Bell Atlantic-NY does not charge extra for the trunks riding the T1. The following is a sample FLEXPATH Switched T1 CSR:

The following explanations refer to Figure 2.7 a-f.

1. *ND6* – Group of 100 numbers. The most common use of a FLEX-PATH T1 is to have 24 DID trunks "ride" the T1. Elements of DID billing previously detailed will also be itemized on the FLEXPATH CSR. The number range is 366-8600 through 366-8699. Cost for the Group of 100 numbers is billed at the back of the CSR.

2. *NGD* – 24 Port Number Group. Bell Atlantic-NY charges $533.06 to provide dial tone to 24 separate channels on the T1.

3. *ND6* - Additional group of 100 numbers. The number range is 366-8700 through 366-8799. Cost for the Group of 100 numbers is billed at the back of the CSR.

4. *NGD* – Additional 24 Port Number Group. Bell Atlantic-NY charges $533.06 to provide 24 separate channels on the T1. Since there are two NGD charges we can deduce that there must be 48 trunks total.

Figure 2.7a

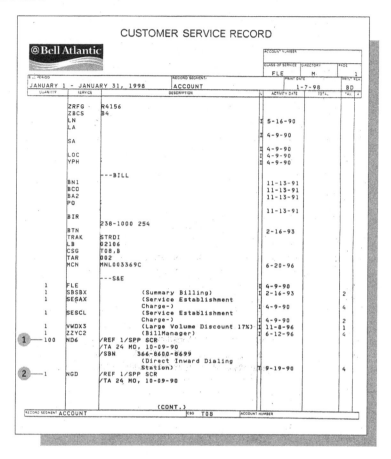

CUSTOMER SERVICE RECORD

Bell Atlantic

ACCOUNT NUMBER		
CLASS OF SERVICE	DIRECTORY	PAGE
FLE	M	1

BILL PERIOD	RECORD SEGMENT	PRINT DATE	DAILY BILL
JANUARY 1 - JANUARY 31, 1998	ACCOUNT	1-7-98	BD

QUANTITY	SERVICE	DESCRIPTION	L	ACTIVITY DATE	TOTAL	TAX
	ZRFG	R4156				
	ZBCS	B4				
	LN		I	5-16-90		
	LA					
	SA		I	4-9-90		
			I	4-9-90		
	LOC		I	4-9-90		
	YPH		I	4-9-90		
		---BILL				
	BN1			11-13-91		
	BCO			11-13-91		
	BA2			11-13-91		
	PO					
	BIR			11-13-91		
		238-1000 254				
	BTN			2-16-93		
	TRAK	STRDI				
	LB	02106				
	CSG	T08,B				
	TAR	002				
	MCN	MNL003369C		6-20-96		
		---S&E				
1	FLE		I	4-9-90		
1	SBSBX	(Summary Billing)	I	2-16-93	2	
1	SESAX	(Service Establishment Charge-)	I	4-9-90	4	
1	SESCL	(Service Establishment Charge-)	I	4-9-90	2	
1	VWDX3	(Large Volume Discount 17%)	I	11-8-96	1	
1	ZZYC2	(BillManager)	I	6-12-96	4	
100	ND6	/REF 1/SPP SCR /TA 24 MO, 10-09-90 /SBN 366-8600-8699 (Direct Inward Dialing Station)	I	9-19-90	4	
1	NGD	/REF 1/SPP SCR /TA 24 MO, 10-09-90				

(CONT.)

RECORD SEGMENT ACCOUNT	CSG T08	ACCOUNT NUMBER

5. *ND6* – Additional group of 100 numbers. The number range is 366-8800 through 366-8899. Right below this ND6 is another ND6 charge for numbers 366-8900 through 366-8999. Costs for each Group of 100 numbers are billed at the back of the CSR

6. The TW6 is the USOC for a DID trunk riding a Switched T1. The trunks are provided with a numbering sequence of 366-8600 001 through 366-8600 048.

Figure 2.7b

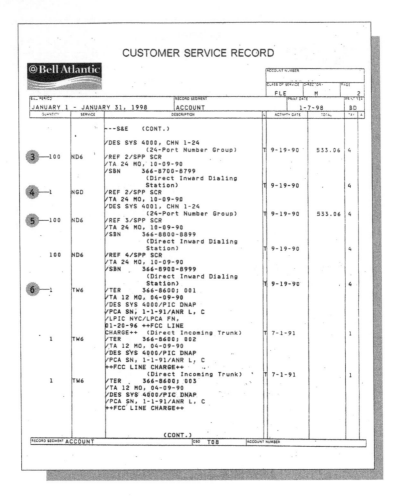

CUSTOMER SERVICE RECORD

Bell Atlantic

			ACCOUNT NUMBER		
			CLASS OF SERVICE	DIRECTORY	PAGE
			FLE	M	2

BILL PERIOD		RECORD SEGMENT		PRINT DATE	PRINT YES
JANUARY 1 - JANUARY 31, 1998		ACCOUNT		1-7-98	BD

QUANTITY	SERVICE	DESCRIPTION	L	ACTIVITY DATE	TOTAL	TAX
		---S&E (CONT.)				
		/DES SYS 4000, CHN 1-24				
		(24-Port Number Group)	T	9-19-90	533.06	4
100	ND6	/REF 2/SPP SCR				
		/TA 24 MO, 10-09-90				
		/SBN 366-8700-8799				
		(Direct Inward Dialing				
		Station)	T	9-19-90		4
1	NGD	/REF 2/SPP SCR				
		/TA 24 MO, 10-09-90				
		/DES SYS 4001, CHN 1-24				
		(24-Port Number Group)	T	9-19-90	533.06	4
100	ND6	/REF 3/SPP SCR				
		/TA 24 MO, 10-09-90				
		/SBN 366-8800-8899				
		(Direct Inward Dialing				
		Station)	T	9-19-90		4
100	ND6	/REF 4/SPP SCR				
		/TA 24 MO, 10-09-90				
		/SBN 366-8900-8999				
		(Direct Inward Dialing				
		Station)	T	9-19-90		4
1	TW6	/TER 366-8600; 001				
		/TA 12 MO, 04-09-90				
		/DES SYS 4000/PIC DNAP				
		/PCA SN, 1-1-91/ANR L, C				
		/LPIC NYC/LPCA FN,				
		01-20-96 ++FCC LINE				
		CHARGE++ (Direct Incoming Trunk)	T	7-1-91		1
1	TW6	/TER 366-8600; 002				
		/TA 12 MO, 04-09-90				
		/DES SYS 4000/PIC DNAP				
		/PCA SN, 1-1-91/ANR L, C				
		++FCC LINE CHARGE++				
		(Direct Incoming Trunk)	T	7-1-91		1
1	TW6	/TER 366-8600; 003				
		/TA 12 MO, 04-09-90				
		/DES SYS 4000/PIC DNAP				
		/PCA SN, 1-1-91/ANR L, C				
		++FCC LINE CHARGE++				

(CONT.)

RECORD SEGMENT ACCOUNT		CSO T08	ACCOUNT NUMBER

7. TBAOX charge is for those trunks that are outbound only as opposed to the TW6 trunks that are inbound only (DIDs).

8. *HCABL* – Since we have 48 trunks we need two T1s. Each T1 needs to have a way for it to be identified (a more detailed explanation is provided in the next chapter). If you are calling the telephone company's repair department you would identify the first T1 as 96,HNNC,2874,,NY.

Figure 2.7c

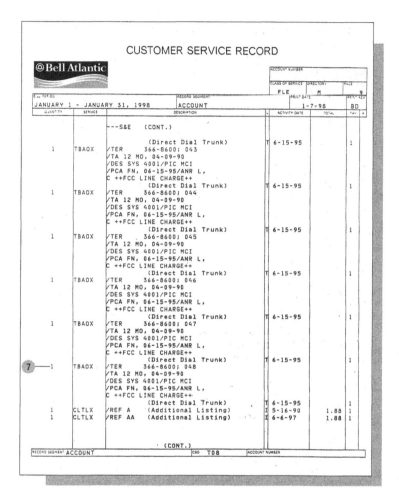

CUSTOMER SERVICE RECORD

Bell Atlantic

			ACCOUNT NUMBER		
			CLASS OF SERVICE	DIRECTORY	PAGE
			FLE	M	9
JANUARY 1 - JANUARY 31, 1998		ACCOUNT		PRINT DATE 1-7-98	BD
QUANTITY	SERVICE	DESCRIPTION	L	ACTIVITY DATE	TOTAL	TAV	A

```
                    └---S&E   (CONT.)

                         (Direct Dial Trunk)        T 6-15-95              1
    1      TBAOX    ✓TER     366-8600; 043
                   ✓TA 12 MO, 04-09-90
                   ✓DES SYS 4001/PIC MCI
                   ✓PCA FN, 06-15-95/ANR L,
                   C ++FCC LINE CHARGE++
                         (Direct Dial Trunk)        T 6-15-95              1
    1      TBAOX    ✓TER     366-8600; 044
                   ✓TA 12 MO, 04-09-90
                   ✓DES SYS 4001/PIC MCI
                   ✓PCA FN, 06-15-95/ANR L,
                   C ++FCC LINE CHARGE++
                         (Direct Dial Trunk)        T 6-15-95              1
    1      TBAOX    ✓TER     366-8600; 045
                   ✓TA 12 MO, 04-09-90
                   ✓DES SYS 4001/PIC MCI
                   ✓PCA FN, 06-15-95/ANR L,
                   C ++FCC LINE CHARGE++
                         (Direct Dial Trunk)        T 6-15-95              1
    1      TBAOX    ✓TER     366-8600; 046
                   ✓TA 12 MO, 04-09-90
                   ✓DES SYS 4001/PIC MCI
                   ✓PCA FN, 06-15-95/ANR L,
                   C ++FCC LINE CHARGE++
                         (Direct Dial Trunk)        T 6-15-95              1
    1      TBAOX    ✓TER     366-8600; 047
                   ✓TA 12 MO, 04-09-90
                   ✓DES SYS 4001/PIC MCI
                   ✓PCA FN, 06-15-95/ANR L,
                   C ++FCC LINE CHARGE++
                         (Direct Dial Trunk)        T 6-15-95              1
 ⑦ ──1     TBAOX    ✓TER     366-8600; 048
                   ✓TA 12 MO, 04-09-90
                   ✓DES SYS 4001/PIC MCI
                   ✓PCA FN, 06-15-95/ANR L,
                   C ++FCC LINE CHARGE++
                         (Direct Dial Trunk)        T 6-15-95              1
    1      CLTLX    ✓REF A   (Additional Listing)    I 5-16-90      1.88    1
    1      CLTLX    ✓REF AA  (Additional Listing)    I 6-6-97       1.88    1


                         (CONT.)
```

RECORD SEGMENT ACCOUNT		CSG T08	ACCOUNT NUMBER

9. *1LDPZ* – This charge is for initial or first local T1 loop ordered between the customer's premise and the customer's local CO. Charge is $435.87.

10. *HCABL* – The second T1 is identified as 96,HNNC,2875,,NY on the CSR.

Figure 2.7d

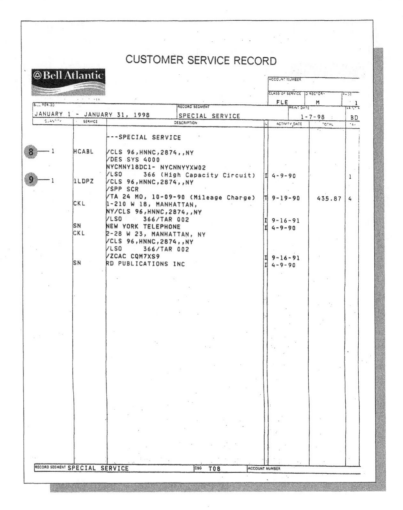

CUSTOMER SERVICE RECORD

Bell Atlantic

			CLASS OF SERVICE	SECTOR	
			FLE	M	1
JANUARY 1 - JANUARY 31, 1998		SPECIAL SERVICE	PRINT DATE 1-7-98		BD
SERVICE		DESCRIPTION	ACTIVITY DATE	TOTAL	

```
                    ---SPECIAL SERVICE
(8)—1    HCABL    /CLS 96,HNNC,2874,,NY
                  /DES SYS 4000
                  NYCMNY18DC1- NYCNNYYXW02
                  /LSO    366 (High Capacity Circuit)   I 4-9-90              1
(9)—1    1LDPZ    /CLS 96,HNNC,2874,,NY
                  /SPP SCR
                  /TA 24 MO, 10-09-90 (Mileage Charge)   T 9-19-90    435.87   4
         CKL      1-210 W 18, MANHATTAN,
                  NY/CLS 96,HNNC,2874,,NY
                  /LSO    366/TAR 002                    I 9-16-91
         SN       NEW YORK TELEPHONE                     I 4-9-90
         CKL      2-28 W 23, MANHATTAN, NY
                  /CLS 96,HNNC,2874,,NY
                  /LSO    366/TAR 002
                  /ZCAC CQM7XS9                          I 9-16-91
         SN       RD PUBLICATIONS INC                    I 4-9-90
```

RECORD SEGMENT SPECIAL SERVICE   CSG T08   ACCOUNT NUMBER

---

11. *1LDP1* – This charge is for each additional T1 local loop between the customer's premise and the customer's local CO. Notice that the price drops from $435.87 to $392.69.

12. We identified four separate ND6 (Group of 100 numbers) charges listed at the beginning of the CSR (366-8600 through 366-8999). The cost is $18.23 per 100 or $72.92 for 400.

Figure 2.7e

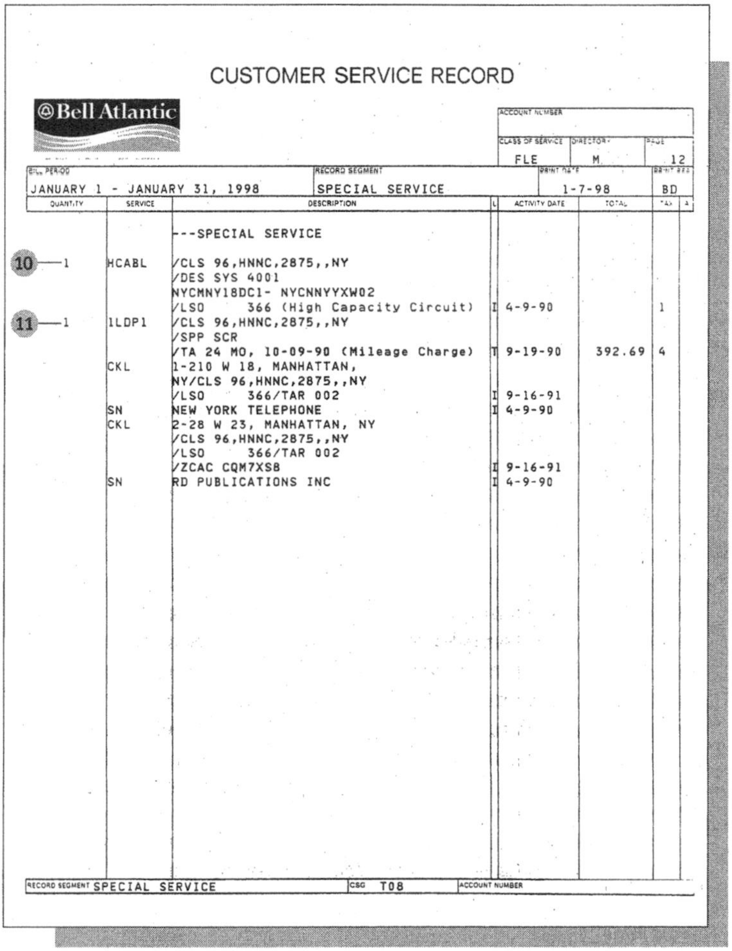

CUSTOMER SERVICE RECORD

Bell Atlantic

	ACCOUNT NUMBER		
	CLASS OF SERVICE	DIRECTOR	PAGE
	FLE	M	12

BILL PERIOD		RECORD SEGMENT		PRINT DATE		PRINT REP
JANUARY 1 - JANUARY 31, 1998		SPECIAL SERVICE		1-7-98		BD

QUANTITY	SERVICE	DESCRIPTION	L	ACTIVITY DATE	TOTAL	TAX	A
		---SPECIAL SERVICE					
⑩—1	HCABL	/CLS 96,HNNC,2875,,NY					
		/DES SYS 4001					
		NYCMNY18DC1- NYCNNYYXW02					
		/LSO 366 (High Capacity Circuit)	I	4-9-90		1	
⑪—1	1LDP1	/CLS 96,HNNC,2875,,NY					
		/SPP SCR					
		/TA 24 MO, 10-09-90 (Mileage Charge)	T	9-19-90	392.69	4	
	CKL	1-210 W 18, MANHATTAN,					
		NY/CLS 96,HNNC,2875,,NY					
		/LSO 366/TAR 002	I	9-16-91			
	SN	NEW YORK TELEPHONE	I	4-9-90			
	CKL	2-28 W 23, MANHATTAN, NY					
		/CLS 96,HNNC,2875,,NY					
		/LSO 366/TAR 002					
		/ZCAC CQM7XS8	I	9-16-91			
	SN	RD PUBLICATIONS INC	I	4-9-90			

RECORD SEGMENT	SPECIAL SERVICE		CSG	T08		ACCOUNT NUMBER	

13. *9ZR* – Each Switched T1 has 24 channels and is charged an 9ZR for each channel. In this example we have two Switched T1s so the total number of 9ZRs billed equals 48.

Figure 2.7f

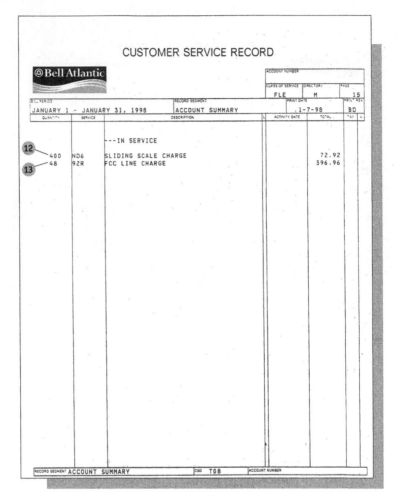

CUSTOMER SERVICE RECORD

Bell Atlantic

			ACCOUNT NUMBER			
			CLASS OF SERVICE	DIRECTORY	PAGE	
			FLE	M	15	
BILL PERIOD		RECORD SEGMENT		PRINT DATE	PRD. AREA	
JANUARY 1 - JANUARY 31, 1998		ACCOUNT SUMMARY		1-7-98	BD	
QUANTITY	SERVICE	DESCRIPTION	L	ACTIVITY DATE	TOTAL	TAY +

```
             ---IN SERVICE

⑫  400    ND6   SLIDING SCALE CHARGE                    72.92
⑬   48    9ZR   FCC LINE CHARGE                        396.96
```

RECORD SEGMENT ACCOUNT SUMMARY	CSG T08	ACCOUNT NUMBER

Dedicated Services

Special services can be broadly defined as any service provided by the Telephone Company that is not POTS. Examples of special service circuits include switched services such as Direct Inward Dial (DID) trunks and Switched T1s. Special service orders often require custom engineering and design work on the part of the telephone company.

A common special service offering is a dedicated point-to-point voice or data line, often referred to as a private line. It is called a private line because you have the exclusive use of it. Most times it is called a circuit to differentiate it from a POTS line. A dedicated circuit bypasses the public switched network. With a POTS line, each time you call someone, a new connection is temporarily established via a CO switch (i.e. a DMS100 or 5ESS) to the called party. The connection lasts as long as the call. You are billed for usage based on the distance between locations and the time the connection stays in place.

Figure 2.8 Dedicated Access

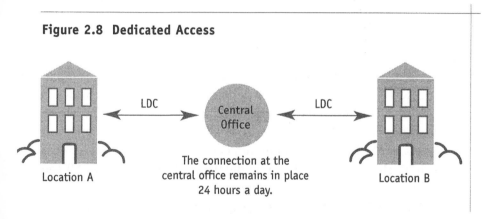

The connection at the
central office remains in place
24 hours a day.

Location A

Location B

A dedicated line also connects two locations. The connection however, remains in place at all times. You pay a higher monthly fee for this type of connection, and for mileage between locations, but you are not billed a separate usage charge. A private line can be used to transmit voice or data. If the line is used for voice services, it allows a company to make unlimited calls between locations at no additional charge. Companies that have multiple offices within a certain geographic area often utilize private lines. The more these offices call each other, the greater the justification for a private line. Companies can also use private lines to connect a mainframe computer at one location with multiple computer terminals or PCs at other locations.

The terms dedicated line; private line and point-to-point circuit are often used interchangeably. There are, however, many different types of data and voice lines that have various technological and pricing differences. There are two main types of dedicated circuits; analog and digital.

Analog circuits are either configured as 2 wire or 4 wire. A 4 wire curcuit can support simultaneous transmission from either location. An analog circuit is often called a voice grade circuit because it operates in the frequency range that best carries voice conversations (300 to 3000 Hertz). An analog circuit can transmit data traffic with the use of a modem. Modems convert data from your PC into a sine wave capable of being transmitted over analog lines. Digital circuits can transmit data without the use of a modem. Conversely, voice conversations must be digitized in order to be sent over digital circuits. Examples of analog circuits are low speed data circuits such as burglar or fire alarms, PBX off-premise extensions and tie lines between PBXs.

Digital circuits include high-speed circuits such as DSO, DS1 and DS3 circuits. DS0 is digital service, level zero and provides bandwidth at 56K or 64K. DS1 is digital service, level one and provides bandwidth at 1.544 MBPS. A DS1 provides the equivalent of 24 DS0s. DS3 is digital service, level three. It provides bandwidth at 44.736 MBPS and is equivalent to 28 DS1s or 672 DS0s. Digital circuits can transmit either data or digital voice. Each voice channel can be digitized into one DS0. A DS3 can carry 672 digitized voice conversations simultaneously.

Billing for dedicated services usually appears at the end of a CSR. Some telephone companies bill special service circuits separately from your regular phone bill on what is known as a special bill. The BTN for a special bill is either the circuit number or a fictitious telephone number (such as 212 S89-1234) that has been assigned to it. If you overlook reviewing your CSR, you could end up paying for disconnected private lines.

Identifying and Understanding Dedicated Circuits

The billing of a dedicated circuit is illustrated in figure 2.9. Figure 2.10 illustrates how a dedicated voice line from a PBX located in one location connected to another location will appear on a CSR. The following items are associated with dedicated circuits and will frequently appear on a CSR:

1. Circuit ID
2. Interoffice Mileage (if applicable)
3. Local Distribution Channels (LDC)
4. Transmission & Signaling Requirements (if applicable)

1. *Circuit ID* - Instead of a standard ten-digit telephone number, special service circuits have an alphanumeric identification. This identification is called the Circuit ID. In example 2.10 the Circuit ID is 96,OSNA,448589. All telephone companies use a standard format to identify circuits. This naming convention is used throughout the industry and is known as Bell System Common Language.

Figure 2.9 Private Line Circuit

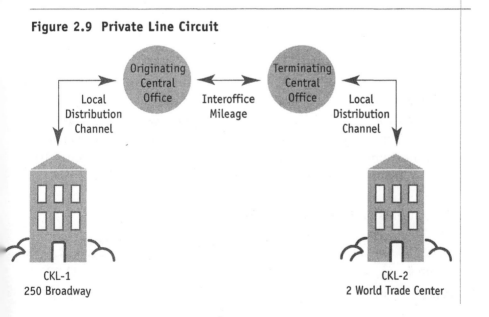

CKL-1
250 Broadway

CKL-2
2 World Trade Center

Figure 2.10

CUSTOMER SERVICE RECORD

@Bell Atlantic

<table>
<tr><td colspan="3"></td><td colspan="3">ACCOUNT NUMBER</td></tr>
<tr><td colspan="3"></td><td>CLASS OF SERVICE
☐F M B X A</td><td>DIRECTORY
X</td><td>PAGE
1</td></tr>
<tr><td>BILL PERIOD
JANUARY 16 - FEBRUARY 15,1999</td><td colspan="2">RECORD SEGMENT
SPECIAL SERVICE</td><td colspan="2">PRINT DATE
1-20-99</td><td>PRINT REA
B D</td></tr>
</table>

QUANTITY	SERVICE	DESCRIPTION	L	ACTIVITY DATE	TOTAL	TAX	A
		---SPECIAL SERVICE					
1	PFSAS	/CLS 96,OSNA,448589,					
		(Private Line Facilities)	I	5-5-92		2	
10	3LN2Y	/CLS 96,OSNA,448589,	D				
		(Mileage Charge)	I	5-5-92	85.68	2	
	CKL						
		/CLS 96,OSNA,448589,					
		/LOC FLR BSM/LSO 718 430					
		/TAR 001	I	5-5-92			
	SN	SM SUB	I	5-5-92			
1	PMWGX	(Type C-Off Premises					
		Extension)	I	5-5-92	6.57	2	
1	CON2X	(Off Premise Extension)	I	5-5-92	21.53	2	
	CKL						
		/CLS 96,OSNA,448589,					
		/DES LEVEL G/LOC RM					
		TELCO/LSO 718 515					
		/TAR 001/ZCAC SPA2ZV4	I	5-5-92			
	SN	SM SUB	I	5-5-92			
1	PMWGX	(Type C-Off Premises					
		Extension)	I	5-5-92	6.57	2	
1	CON2X	(Off Premise Extension)	I	5-5-92	21.53	2	

The Circuit ID can either be in telephone number format or in serial number format. The /CLS FID after a circuit indicates the Circuit ID is in serial number format, while the /CLT FID identifies a Circuit ID that is in telephone number format. In our example below the Circuit ID 96,OSNA,448589,,NY is in serial number format.

A Circuit ID is divided into sections as follows:

	Prefix		Service Code		Modifier		Serial Number					
Position #	1	2	3	4	5	6	7	8	9	10	11	12
Character	9	6	0	S	N	A	4	4	8	5	8	9

Telecommunications Expense Management

The number 96 (positions 1 & 2) and the serial number (position 7 through 12) simply allow the telephone company to assign a unique number to each circuit. Positions 3 & 4 make up what is known as the service code. The service code identifies both the type of circuit and how the circuit is used. When taking your order, a telephone company representative must assign the correct service code to your request. This allows the telephone company's engineering department to design the circuit according to your needs.

Each special service circuit fulfills unique needs and is charged accordingly. Therefore, it is critical for the telephone company representative to determine the proper code for your needs and requirements. The following chart lists common service codes:

Common IntraLATA Special Circuit Service Codes

Service Code	Description
AB	PACKET SWITCH TRUNK
AD	ATTENDANT
AF	COMMERICAL AUDIO (FULL TIME)
AI	AUTOMATIC OUTWARD DIALING
AL	ALTERNATE SERVICES
AM	PACKET OFF NETWORK ACCESS LINE
AN	ANNOUNCEMENT SERVICE
AO	INTERNATIONAL/OVERSEAS AUDIO (FULL TIME)
AP	COMMERICAL AUDIO (PART TIME)
AT	INTERNATIONAL/OVERSEAS AUDIO (PART TIME)
AU	AUTO SCRIPT
BA	PROTECTIVE ALARM (DC)
BL	BELL & LIGHTS
BS	SIREN CONTROL
CA	SSN ACCESS
CB	OCC AUDIO FACILITIES

Service Code	Description
CC	OCC DIGITAL FACILITY-MEDIUM SPEED
CE	SSN STATION LINE
CF	OCC SPECIAL FACILITY
CG	OCC TELEGRAPH FACILITY
CH	OCC DIGITAL FACILITY- HIGH SPEED
CI	CONCENTRATOR IDENTIFER TRUNK
CJ	OCC CONTROL FACILITY
CK	OCC OVERSEAS CONNECTING FACILITY WIDEBAND
CL	CENTREX CO LINE
CM	OCC VIDEO FACILITY
CN	SSN NETWORK TRUNK
CO	OCC OVERSEAS CONNECTING FACILITY
CP	CONCENTRATOR IDENTIFIER SIGNALING TRUNK
CR	OCC BACKUP FACILITY
CS	CHANNEL SERVICE (UNDER 300 BAUD)
CT	TIE TRUNK
CV	OCC VOICE GRADE FACILITY
CW	OCC WIRE PAIR FACILITY
CX	CENTREX CU STATION LINE
CZ	OCC ACCESS FACILITY
DA	DIGITAL DATA OFF-NET EXTENSION
DB	HSSDS 1.5 MBPS/s ACCESS LINE
DF	HSSDS 1.5 MBPS/s HUB TO HUB
DG	HSSDS 1.5 MBPS/s HUB TO EARTH STATION
DH	DIGITAL SERVICE - 1.544 Mbps
DI	DIRECT DIAL IN TRUNK
DJ	DIGITAL TRUNK

Service Code	Description
DK	DATA LINK
DL	DICTATION LINE
DO	DIRECT DIAL-OUT
DS	SUBRATE SPEEDS - DIGITAL DATA
DW	DIGITAL DATA - 56 KBS
EA	SWITCHED ACCESS
EB	END OFFICE TRUNK
EC	TANDEM TRUNK
EE	COMBINED ACCESS
EF	ENTRANCE FACILITY-VOICE GRADE
EG	TYPE #2 TELEGRAPH
EN	EXCHANGE NETWORK ACCESS FACILITY
EP	ENTRANCE FACILITY-PROGRAM GRADE
ES	EXTENSION SERVICE VOICE GRADE
ET	ENTRANCE FACILITY-TELEGRAPH GRADE
EU	EXTENSION SERVICE -TELEGRAPH GRADE
EW	OFF-NETWORK MTS/WATS EQUIVALENT
FD	PRIVATE LINE-DATA
FG	GROUP SPECTRUM
FR	FIRE DISPATCH
FT	FOREIGN EXCHANGE TRUNK LINE
FW	WIDEBAND CHANNEL
FV	VOICEGRADE FACILITY
FX	FOREIGN EXCHANGE LINE
HW	DIGIPATH II
IT	INTERTANDEM TIE TRUNK
LA	LOCAL AREA DATA CHANNEL

Service Code	Description
LL	LONG DISTANCE TERMINAL LINE
LS	LOCAL SERVICE
LT	LONG DISTANCE TERMINAL TRUNK
MA	CELLULAR ACCESS TRUNK 2-WAY
MT	WIRED MUSIC
ND	NETWORK DATA LINE
OC	CENTREX CU STATION LINE OFF-PREMISES
OI	OFF PREMISES INTERCOMMUNICATION LINE
ON	OFF-NETWORK ACCESS LINE
OP	OFF-PREMISES EXTENSION
OS	OFF-PREMISES PBX STATION LINE
PA	PROTECTIVE ALARM (AC)
PC	SWITCHED DIGITAL ACCESS LINE
PG	PAGING
PL	PRIVATE LINE - VOICE
PM	PROTECTIVE MONITORING
PP	PICTURE PHONE LINE
PX	PBX STATION LINE
QS	PACKET ACCESS LINE
RA	REMOTE ATTENDANT
RT	RADIO LAND LINE
SA	SATELLITE TIE TRUNK
SG	REMOTE METERING-SIGNAL GRADE
SL	SECRETARIAL LINE
SM	SAMPLING
SN	SSN-SPECIAL ACCESS TERMINATION
SS	DATAPHONE

Service Code	Description
TA	TANDEM TIE TRUNK
TC	REMOTE MONITORING TELEGRAPH GRADE
TF	TELEPHOTO/FACSIMILE
TK	LOCAL PBX TRUNK
TL	NON-TANDEM TIE TRUNK
TR	TURRET OR AUTOMATIC CALL DISTRIBUTOR (ACD) TRUNK
TT	TELETYPEWRITER CHANNEL
TU	TURRET OR AUTOMATIC CALL DISTRIBUTOR (ACD) LINE
VF	COMMERICAL TELEVISION (FULL TIME)
VH	COMMERICAL TELEVISION (PART TIME)
VM	CONTROL/REMOTE METERING VOICE GRADE
VO	INTERNATIONAL/OVERSEAS TELEVISION
VR	NON-COMMERICAL TELEVISION
WC	SPECIAL 800 SERVICE TRUNK
WD	SPECIAL WATS TRUNK OUT
WI	SPECIAL 800 SERVICE OUT
WO	WATS LINE OUT
WS	WATS TRUNK OUT
WX	WATS TRUNK 2-WAY
WZ	WATS LINE 2-WAY
ZA	ALARM CIRCUITS

To find out what the service code "OS" in our sample Circuit ID signifies, look at the preceding list. The service code OS is an off premise PBX station line. This customer has a PBX and a dedicated circuit to a satellite location. The telephone at the satellite location operates as an extension off the PBX.

A tie line (a private line that connects two PBXs) would be identified by the use of the service code TL. The Circuit ID would then read 96,TLNA,448589.

Position 5 of the Circuit ID tells you the intended use of the circuit. Position 5 can be either A,D, N or S. Each alpha character means the following:

A – Alternates between data and non-data

D – Purely data

N – Purely non-data

S – Simultaneous data and non-data

In our original example, Circuit ID 96,OSNA,448589,,NY, the 5th character is an "N", so we know the circuit is an off-premise extension from the "OS" and is used for non-data purposes.

Character 6 covers who provides certain facilities, the local telephone company, the long distance carrier or your equipment vendor. The "NY" at the end of the Circuit ID tells us the circuit is operational in New York.

A Circuit ID listed in telephone number format would appear as follows: 96,DINA,212,555,7400,001. The DI indicates it is for a DID trunk.

The format for a circuit in telephone number format is listed below:

	Prefix		Service		Modifier		NPA			CO Line#							Ext.		
Position #	1	2	3	4	5	6	7	8	9	10	11	12	13	14	15	16	17	18	19
Character	9	6	D	1	N	A	2	1	2	5	5	5	7	4	0	0	0	0	1

2. *Interoffice Mileage* - You pay more for a circuit that extends across town than you do for one connecting locations on different floors of the same building. Mileage billing contains two major components. The first component is a fixed minimum charge that is always billed once a circuit connects locations served by different COs regardless of

the distance between COs. The only way to find mileage errors is to purchase a software package that calculates the mileage based on the NPA/NXX of both locations.

The second component is known as interoffice mileage charges (INOF in telephone jargon) and it is strictly based on the distance between circuit connections. You are billed for the actual air miles between the telephone company CO that serves your location and the Telephone Company CO that provides service to the terminating end of the circuit (sometimes called the foreign location).

3. *Local Distribution Charge (LDC)* - The loop charge covers the cost of the connection between each location and its local CO. A circuit will have two LDC charges, one for each end.

4. *Transmission & Signaling* - Transmission packages are based on the type of equipment that is connected to the circuit. Different design parameters are required from the Telephone Company depending on how and for what purpose the circuit will be utilized. Your equipment vendor specifies what type of transmission is required.

Signaling determines how one location of a circuit tells the other location that they are calling or sending data. It determines, for example, if a telephone will ring automatically at one location if someone picks up a phone at the originating location. Bell Atlantic-NY offers the following transmission options called Feature Functions:

Feature Function	Monthly Rate
Basic 2-Wire Voice	$ 7.07
Basic 4-Wire Voice	$19.99
Type A PXOS Feature	$12.77
Type B PXOS Feature	$12.77
Type C PXOS Feature	$ 6.57
Basic 2-Wire Data	$ 4.67
Basic 4-Wire Data	$11.65
Enhanced 4-Wire Data	
-Type C-1	$11.65
-Type C-2	$11.65
-Type C-4	$11.65
-Type D-1	$11.65

Armed with a basic knowledge of USOCs and the major components of a special service circuit, you can now analyze a CSR that contains billing for a special service circuit. The following explains the billing for the special service portion of a Bell Atlantic-NY CSR.

Figure 2.11

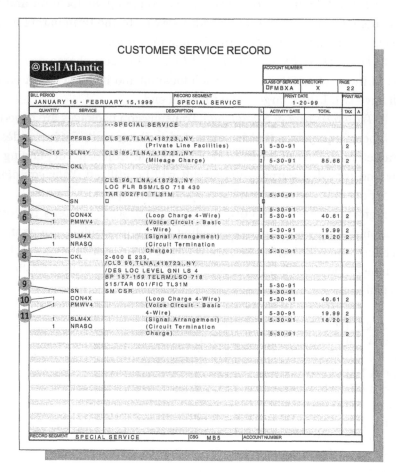

1. *PFSBS* - This USOC identifies the circuit ID as 96,TLNA,418723,,NY. We can tell it is in serial number format from the CLS FID. This customer has a tie line that connects two PBXs at two different locations. We know this from service code TL (see preceding list of codes).

2. *3LN4Y* - Prefixes like 3LN, 1LN or 1LB indicate interoffice mileage. When the second character of the USOC is an "L", you can be sure that the USOC generates mileage billing. The USOC "4Y" at the end tells us that this is a 4-wire circuit. Every telephone company bills at different rates. Rates can be obtained by checking the telephone company tariffs or by calling the telephone company directly. The number ten (10) in the quantity field indicates that the distance between the central office serving the originating location and the central office serving the terminating location is 10 (1/4) miles or 2 1/2 miles. The mileage charge between central offices is determined by calculating the distance in a trajectory akin to airline miles using either Latitude and Longitude or Vertical and Horizontal (V&H) coordinates. Telephone company personnel will often use the term "as the crow flies" (meaning a straight line) to describe the calculation of mileage between COs. The mileage is calculated by entering the LSOs (which are the NPA and NXX of each location) into a program that calculates mileage. As is noted below, the LSOs are 718 430 for CKL 1 and 718 515 for CKL 2.

3. *CKL* - USOC denoting the circuit location of the "1" or originating location. In this example, CKL-1 is 250 Broadway. The /LSO 718 430 indicates the NPA is 718 and the NXX is 430 for this location.

4. *SN* - This USOC denotes the listed (in the white pages) customer at the CKL 1 address. In this case we have removed the customer name.

5. *CON4X* - USOC denoting a loop charge for a 4-wire circuit (CON2X would be the USOC for a 2-wire circuit). A 4-wire circuit permits simultaneous transmission between locations. This charge is for the LDC for the local loop between the originating location and its serving CO.

6. *PMWV4* - This USOC allows the circuit to be configured to carry voice at the originating end.

7. *SLM4X* – Specialized signaling ordered by the PBX vendor at the originating end.

8. *CKL 2* - Refers to the distant "2" or terminating location of the circuit. This circuit terminates at 600 E. 233 and its LSO is 718 515.

9. *SN* - USOC that denotes the name of the customer at the terminating end. In this case "sm csr" means Same Customer. This tells us the listed name of the company at the terminating end is the same as the listed name on the CSR.

10. *CON4X* - LDC charge for the terminating end (CKL 2).

11. *PMWV4* – This USOC allows the circuit to be configured to carry voice at the terminating end.

Schematically this circuit is depicted as follows:

Figure 2.12 4-Wire PBX Tie Line

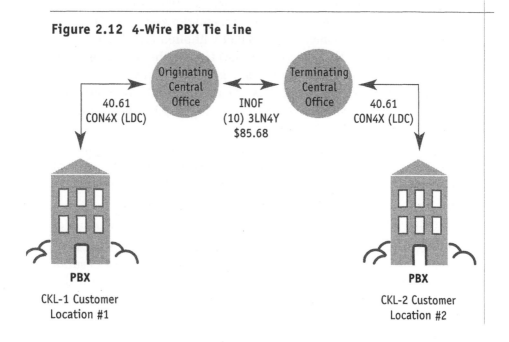

PBX

CKL-1 Customer
Location #1

PBX

CKL-2 Customer
Location #2

Bell Atlantic-NJ CSR

Some telephone companies identify the originating and terminating (CKL 1 and CKL 2) locations at the beginning of the CSR within the LIST section. The CKL will appear after the LOC FID.

Each CKL location is usually assigned a letter of the alphabet. When the Circuit ID is listed in the S&E section, the CKL FID will have two alphas after it. You must refer back to the LIST section of the CSR to determine the originating and terminating ends of the circuit. The first alpha refers to the originating location; the second denotes the terminating location. The following example illustrates how the locations of this circuit are identified.

When you review this Bell Atlantic-NJ CSR, you'll notice slight Circuit ID naming variations. In the previous example OSNA 999293NJ is the Circuit ID. The Circuit ID is not prefixed with a number as is the case with New York circuits, nor do commas separate the components of the Circuit ID.

1. ARR - Lists the originating and terminating locations. CKL (A) is located at 27 Port Jersey Blvd, in the Guardhouse while CKL (P) is at l7 Terrace Drive, Upper Saddle River.

2 & 3. LDC charge - The charge for the portion of your circuit that connects the CO and each location.

4. Mileage component - As with Bell Atlantic-NY, Bell Atlantic-NJ charges mileage based on the distance between the COs serving each end of the circuit. Bell Atlantic-NJ bills mileage at different rates based on the speed and design of the special service circuit. In this example, Bell Atlantic-NJ bills mileage at a rate of $11.67 for the first mile, and $3.35 for each additional mile.

Figure 2.13 Bell Atlantic - NJ

(201) 555-5200		
05-24-91	LN	ABC Corporation
05-24-91	LA	1 Port Jersey Blvd, J CY
05-24-91	SA	302 Port Jersey Blvd, J CY
05-24-91	LOC	Entire Bldg
06-26-86	CKL	(A) 27 Port Jersey Blvd J CY; Guardhouse
02-15-85	CKL	(B) Cargo Area, Port Jersey Blvd, J CY
02-15-85	CKL	(C) Guardhouse #2, Port Jersey Blvd, J CY
02-15-85	CKL	(D) Office, Port Jersey Blvd, J CY
02-15-85	CKL	(E) Annex, Port Jersey Blvd, J CY

11-21-88	CKL	(P) 17 Terrace Drive, Up Sadl RI	
		Service and Equipment Charges	
11-21-88	CKT	OSNA 999293NJ	
(1)	ARR	PX 4XAYS/CKL A, P	
		PVLLS /** PVT LN-Voice Grade- CPE STS EQP	No Charge
(2)		9B1SX /CKL A/LSO 201 555/DES Bldg Guardhouse/LOC Channel DIF EXCH-2014	23.46
(3)		9B1SX /CKL P/LSO 201 999/LOC Channel DIF EXCH-2014	23.46
(4)	25	1LVA4 /SEC /EX J CY, Ramsy/** Mileage	92.07

Schematically our circuit looks like the following:

Figure 2.14

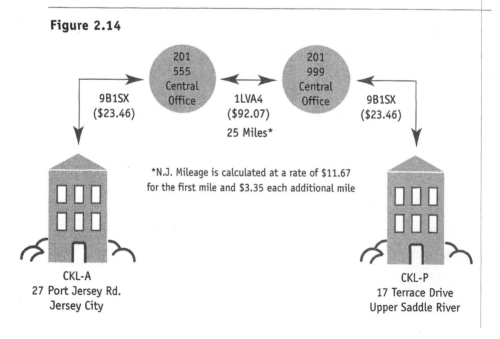

CKL-A
27 Port Jersey Rd.
Jersey City

CKL-P
17 Terrace Drive
Upper Saddle River

Digipath Digital Service II

Digipath, as offered by Bell Atlantic-NY, is a digital circuit that can transmit data at speeds of 2.4, 4.8, 9.6, 19.2, and 56KBS between multiple terminals and/or hosts. A CSR will bill as follows:

1. *DYDRL* - This USOC identifies the Circuit ID as 96,HWDA, 424135,,NY. We can tell it is in serial number format from the CLS FID. The Service Code chart shows HW as Digipath II.

Figure 2.15

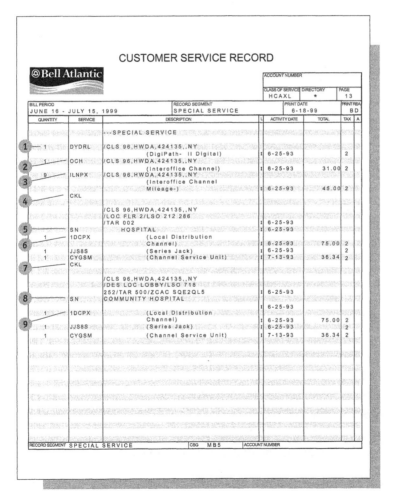

2. *OCH* – This is the USOC for fixed mileage billed at $45.00.

3. *1LNPX* – USOC for variable mileage. The "9" in the quantity field tells us there are 9 miles between the COs serving the originating and the terminating locations.

4. *CKL* - USOC denoting the circuit location of the "1" or originating location. The LSO is 212 286 for this location.

5. *SN* – This USOC denotes the listed (in the white pages) customer at the CKL 1 address. In this case we have removed the customer name.

6. *1DCPX* – LDC for the originating end is billed at $75.00 per end.

7. *CKL 2* – Refers to the distant "2" or terminating location of the circuit. This is where the circuit terminates and the LSO is 718 252.

8. *SN* – USOC that denotes the name of the customer at the terminating end.

9. *1DCPX* – LDC for the terminating end is billed at $75.00 per end.

T1 Billing

A T1 or DS1 when provided as digital service can terminate at the CO, or it can be connected to another location, much like any other special service circuit. If, for example, you have 14 FDDA (full duplex-data circuits) and 10 TLNA (tie lines) circuits to a satellite location across town, these data lines can be consolidated onto a single T1. A T1 that connects two locations (other than a connection to the CO as shown with a Switched T1) is a point-to-point T1.

A T1 is often used to connect voice lines directly to a long distance carrier's POP (point of presence). This arrangement utilizes a point-to-point T1 that originates at your location and terminates at the POP. The Telephone Company billing system called CABS (Carrier Access Billing System) usually bills this type of connection under the FCC tariff. CABS billing is reviewed in Chapter 3.

T1s are billed two LDC charges much like other special service circuits. Some telephone companies add an additional charge based on the distance between your location and your serving CO. (Reminder: Interoffice Mileage is the distance between COs)

Figure 2.16

CUSTOMER SERVICE RECORD

QUANTITY	SERVICE	DESCRIPTION	L	ACTIVITY DATE	TOTAL	TAX	A
		---SPECIAL SERVICE					
1	HCAXS	/CLS 96,DHZA,417327,,NY (High Capacity Circuit)	I	1-3-96		4	
1	XUW1X	/CLS 96,DHZA,417327,,NY (Mileage Charge)	I	1-3-96	115.00	4	
10	3LN1S	/CLS 96,DHZA,417327,,NY (Milage Charge)	I	1-3-96	300.00	4	
1	CCO	/CLS 96,DHZA,417327,,NY Distribution Channel)	I	1-3-96		4	
	CKL	MANHATTEN, NY /CLS 96,DHZA,417327,,NY /LOC FLR GRD/LSO 212 304 /TAR 002	I	1-3-96			
	SN		I	1-3-96			
1	XUN1X	(Mileage Charge)	I	1-3-96	210.00	4	
1	XUD1Y	(Mileage Charge)	I	1-3-96	5.00	4	
	CKL	NY /CLS 96,DHZA,417327,,NY /LOC FLR 7/LSO 212 228 /TAR 002/ZCAC STB2PD8	I	1-3-96			
	SN		I	1-3-96			
1	XUN1X	(Mileage Charge)	I	1-3-96	210.00	4	
1	XUD1Y	(Mileage Charge)	I	1-3-96	5.00	4	

Account Number — Class of Service: FLE — Directory: M — Page: 10
Bill Period: JUNE 13 - JULY 12, 1999 — Record Segment: SPECIAL SERVICE — Print Date: 6-16-99 — Print Rea: BD
Record Segment: SPECIAL SERVICE — CSG: M C 1 — Account Number

Point-To-Point T1 Billing

Let's analyze a condensed CSR for a T1 from Bell Atlantic-NY. Rates will often vary by areas within a former Bell Company. Bell Atlantic-NY has two sets of T1 rates for what they call "downstate" (the New York City Metropolitan Area) and another set of rates for upstate New York. These rates are intended to reflect the differences in the cost of doing business in these different areas. The following CSR details billing for a T1 within the New York Metropolitan-Area (LATA 132):

1. *HCAXS* – The Circuit ID is 96,DHZA,417327,,NY. – The T1 is given a Circuit ID and can be identified by checking positions 3, 4, and 5 (as with a standard special service circuit). Referring back to the service code list, we can see that DH is the service code for Digital Data Service. The Circuit ID is in serial number format. The following analyzes a T1 Circuit ID:

2. *XUW1X* - USOC that denotes the Interoffice Channel Charge. This charge represents the fixed mileage charge regardless of the distance between the COs serving each location.

3. *3LN1S* - USOC that denotes the Variable Interoffice Mileage Charge. The quantity field tells us that the INOF mileage is 10 miles.

4. *CKL1* - This USOC denotes the originating location.

5. *SN* – USOC that denotes the name of the customer at the originating location.

6. *XUN1X* - USOC that denotes the fixed Local Distribution Channel charge. The Local Distribution Channel (LDC) is the channel from your local CO to your premise.

7. *XUD1Y* - USOC that denotes the Local Distribution Mileage Charge. This is where you are charged extra for the distance (in 1/4 miles) that you are physically located from the CO.

8. *CKL 2* - The terminating location.

9. *SN* - The name of the customer at the terminating location.

10. & 11. - Repeat of numbers 6 & 7 for the terminating end.

Bell Atlantic-NJ will bill a point-to-point T1 on a CSR as follows (condensed CSR):

	Prefix		Service		Modifier		NPA			CO Line#							Ext.		
Position #	1	2	3	4	5	6	7	8	9	10	11	12	13	14	15	16	17	18	19
Character	9	6	D	1	N	A	2	1	2	5	5	5	7	4	0	0	0	0	1

List Section

LST	ABC HOSPITAL
LA	101 NEW JERSEY TURNPIKE, NWK
CKL	(A) 12 BAYONNE ROAD
CKL B	(B) 234 12TH ST., PORT JERSEY

Service and Equipment Section

Quantity	Item	Description	Rate
1	CKT	DHSA 999041NJ	
1	DIF	1.544 MBPS	
1	HCAXS	/HI CAP CIRCUIT	
1	1ROP2	/CKL A/LSO 201 999	$215.00
1	1ROP2	/CKL B/LSO 201 555	$215.00
6	1LWPX	INTEROFFICE CHANNEL	$292.00

The total cost of a point-to-point T1 in New Jersey that connects locations that are served by COs 6 miles apart is $722.

3

Carrier Access Billing System "CABS"

Access is the means by which customers reach an interexchange carrier (IXC) network and the IXC reaches its customers. It is usually owned, tariffed and controlled by a local exchange carrier (LEC) or other local provider.

CABS (Carrier Access Billing System) is the billing system that was created by the LECs to bill the IXCs for using their telephone lines and dedicated facilities to reach customers. CRIS is the billing system that is used by the LECs to bill end users (residence and business customers).

The services that the LECs bill via their CABS system are regulated by the FCC while services billed by the LEC's CRIS system are regulated by each state's Public Service Commission (P.S.C.) or Public Utilities Commission (P.U.C.). For example, Ameritech Access Services and rates that are billed by Ameritech's CABS system are filed in Ameritech's FCC Tariff 2 (Access Services). The services found in Ameritech's FCC Tariff 2 apply to Indiana, Illinois, Michigan, Ohio, and Wisconsin. The description for the services listed in the FCC Tariff 2 are the same for all of the states even though the pricing may differ from one state to another. Ameritech has five separate state tariffs filed with the PUCs in Michigan, Illinois, Indiana, Ohio and Michigan.

As the CABS system was developed to bill IXCs, the pricing was designed to be at the wholesale level. However, larger corporate customers began to realize that services offered under the FCC tariffs tended to be cheaper. Leased line services (analog and digital) are available under the LEC's FCC tariff as well as under their P.S.C. tariffs. Savvy customers now "tariff shop". They order services under the tariff that provides the best rate.

Uniformity in the LEC's Access Tariff:

The IXCs interface with the LEC by issuing a request for service called an ASR (Access Service Request). The specifications for the ASR are standard throughout the country and can be ordered from Telecordia (formerly Bellcore) by calling 800 521-2673.

As a result of this standardization of the ASR, a CABS CSR is the same across the country. FCC tariffs also generally follow the same format. Switched access services and rates are found in section 6 of the access tariff and special access services and rates are found in Section 7.

Two General Types of Access Services:

Two basic types of access exist, Switched and Dedicated (Special Services).

Switched Access Service: Switched access calls are transmitted partially on common transport circuits or on shared lines. With switched access a customer's call goes to the LEC's CO switch before it is routed on to a long distance carrier. Traffic from COs within a LATA is funneled over shared circuits to a LEC's access tandem switch. These calls are split out by the IXC and transmitted to a designated IXC's point of presence (POP). The LEC bills the IXC for transporting the call from a customer's location to the IXC's POP via the CABS system. These charges are outlined in the LEC's FCC tariff.

Three main types of switched access are available from the LEC:

Feature Group A (FGA) – With FGA, a long distance carrier's POP appears to be simply another telephone to a LEC switch. FGA is considered a "line-side" connection that connects on the side of the switch on which standard telephone lines connect. With FGA, IXC customers must dial a local telephone number to connect to the IXC. A PIN must be entered before a long distance call can be dialed.

Feature Group B (FGB) – FGB is similar to FGA except that it provides a higher quality trunk-side connection between the LEC and the IXC.

FGA and FGB are only used where FGD is not available.

Feature Group D (FGD) – The 1984 divestiture of the Bell System mandated that the RBOCs provide all IXCs access arrangements equal to that offered to AT&T (called equal access). FGD provides equal access to all IXCs where available. FGD can only be provided by an electronic switching center. Most of the RBOC switches have been updated to provide FGD.

With FGD, customers can access their long distance carrier in two ways:

- *Presubscribed access* – A customer chooses an IXC as their primary carrier by notifying their LEC. To complete long distance calls the customer simply has to dial "1", the area code and the telephone number.

- *Non-presubscribed access* – Customers can reach other IXCs by dialing an access code such as 10333 before dialing the long distance call.

The special Access Services tariff section outlines the basic configurations, rate elements and rate structures associated with Special Access Services which provide the transmission path (physical access) between a customer's premises and the LEC's Serving Wire Center. It provides for the interconnection between both an end-user and a Carrier, Carrier and another carrier, or an end-user customer and an end-user customer. Special Access Services can be used for voice or data functions; and are categorized as analog or digital services. Spectrum and bandwidth differentiate analog connections; and digital connections are differentiated by bit rate.

The following diagram depicts a two-point Direct Analog Service connecting two customer designated premises located 15 miles apart. The service is provided with C-Type conditioning.

Figure 3.1

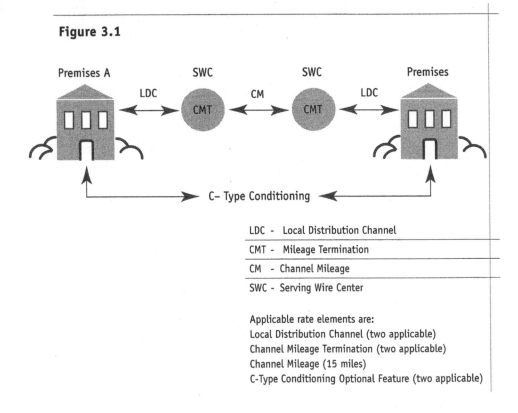

LDC - Local Distribution Channel

CMT - Mileage Termination

CM - Channel Mileage

SWC - Serving Wire Center

Applicable rate elements are:
Local Distribution Channel (two applicable)
Channel Mileage Termination (two applicable)
Channel Mileage (15 miles)
C-Type Conditioning Optional Feature (two applicable)

Rate Categories- There are four basic rate categories that apply to Special Access Services:

(A) *Local Distribution Channel (LDC)* is also called a Channel Termination (CT) by some LECs - provides for the communication path between a customer's designated premises and its Serving Wire Center (SWC). One LDC or CT charge applies per customer designated premises at which the channel is terminated. This charge will also apply even if the customer-designated premises are co-located in a Telephone Company building. For example Telephone Company Centrex CO-like switches are considered to be the customer's premises for the purpose of determining the number of LDCs required to provision a special access circuit.

(B) *Channel Mileage Termination (CMT)* rate category provides for the termination of transmission facilities between the serving wire centers associated with two customers designated locations. Some LECs do not charge this item.

(C) *Channel Mileage (CM)* – Rate category provides for termination of transmission facilities between two SWCs. Many LECs charge both a fixed mileage fee (flat rate anytime the mileage is over zero) and a variable per mile rate.

(D) *Optional Features and Functions* - Rate categories for optional features such as signaling and conditioning. Signaling capability, hubbing functions and conditioning fall into this category.

CABS CSR

The LEC billing center that will investigate CABS billing problems is known as the Interexchange Carrier Service Center (ICSC). The BTN is referred to as the Billing Account Number (BAN) on a CABS bill. Your monthly bill for a CABS account is rendered under what is known as a special bill number. The Bell Atlantic - NY CABS bill format is M51-XXXX while Bell Atlantic – NJ utilizes R15-XXXX. Ameritech – Michigan utilizes E88-XXXX.

The CABS system has its own USOCs and FIDS. Here is a sample of some CABS USOCs and FIDs:

USOC/FID	Definition
ACNA	ACCESS CUSTOMER NAME ABBREVIATION
CKL	CIRCUIT LOCATION
CLS	CIRCUIT ID - SERIAL NUMBER FORMAT
1L5LS	MILEAGE
1RL2W	VOICEGRADE IMPROVED RETURN 2 WIRE
1RL4W	VOICEGRADE IMPROVED 4 WIRE
LAT	LOCAL ACCESS TRANSPORT AREA ID
LOC	LOCATION
LSO	LOCAL SERVICE OFFICE
MPB	MEET POINT BILLING
NC	NETWORK CHANNEL
NCI	NETWORK CHANNEL INTERFACE
PF	PRINT FREQUENCY
PIU	PERCENTAGE OF INTERSTATE USAGE
POI	POINT OF INTERFACE
SED	SERVICE ESTABLISHMENT DATE
SPP	SPECIAL PRICING PLAN
SN	SERVICE NAME
S25	SPECIAL ACCESS SURCHARGE
TA	TERM AGREEMENT
TMECS	CHANNEL TERM – 1.544 MBPS
T6E2X	CHANNEL TERMINATION 2 WIRE
T6E4X	CHANNEL TERMINATION 4 WIRE
UTM	SPECIAL ACCESS RECOVERY SURCHARGE
XDV2X	ANALOG VOICEGRADE 2 WIRE
XDV4X	ANALOG VOICEGRADE 4 WIRE
XSSLR	20HZ RINGDOWN INTERFACE
ZCR	CORRIDOR SERVICE

A CABS Circuit ID will be in serial number format. Just as with intraLATA special service circuits, the Circuit ID can be broken down into its components. A sample Circuit ID of 32,LGGS999123,NY is defined as follows:

	Prefix		Service Code		Modifier		Serial Number					
Position #	1	2	3	4	5	6	7	8	9	10	11	12
Circuit ID	3	2	L	G	G	S	9	9	9	1	2	3

NOTE: the "NY" above indicates that Bell Atlantic - NY, provisioned this circuit.

The service code defines the service the circuit is providing. The following lists common service codes.

interLATA Service Codes (Circuit Position #3 & #4)

USOC/FID	Definition
HC	HIGH CAPACITY 4.544 MBPS
HD	HIGH CAPACITY 3.152 MBPS
HE	HIGH CAPACITY 6.312 MBPS
HF	HIGH CAPACITY 44.736 MBPS
LB	VOICE-NON SWITCHED
LC	VOICE-SWITCHED LINE
LF	LOW SPEED DATA
LG	BASIC TRUNK LINE
LH	TIE TRUNK
LJ	VOICE AND DATA SSN ACCESS
LK	VOICE AND DATA-INTERMACHINE TRUNK
LN	DATA EXTENSION
SB	SWITCHED ACCESS-STANDARD
SD	SWITCHED ACCESS-IMPROVED

USOC/FID	Definition
SE	SPECIAL ACCESS-DEDICATED
SF	SPECIAL ACCESS-DEDICATED IMPROVED
WB	WIDEBAND DIGITAL 19.2 kb/s
WE	WIDEBAND DIGITAL 50 kb/s
WF	WIDEBAND DIGITAL 230.4 kb/s
WH	WIDEBAND DIGITAL 56 kb/s
XA	DEDICATED DIGITAL 2.4 kb/s
XB	DEDICATED DIGITAL 4.8 kb/s
XG	DEDICATED DIGITAL 9.6 kb/s
XH	DEDICATED DIGITAL 56 kb/s

The CABS system also utilizes Common Language Location Identifier (CLLI) Codes. CLLI (pronounced silly) Codes are created and maintained by Telecordia (formerly Bellcore). CLLI Codes are a telecommunications standard for identifying circuit locations and are used both by LECs and IXCs. Every POP and CO is assigned a CLLI code. Many large buildings also have a CLLI code assigned to it. A sample CLLI Code (WHPLNYWPK42) is defined as follows:

Figure 3.2

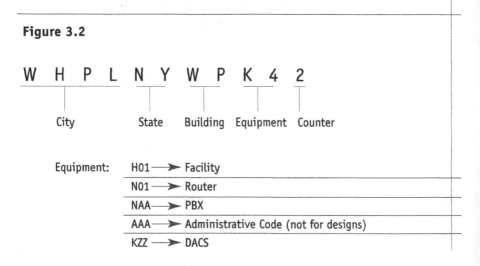

CABS CSRs utilize Common Language Facility Identifiers (CLFI) as well. The CLFI identifies circuits that carry sub-rate circuits within their bandwidth. An example of an CLFI is 5000.T3.11.NYCMNYZRW01.WHPL-NYWPK42.

Element #	1	2	3	4	5
CLFI	5000	T3	11	NYCMNYZRW01	WHPLNYPK42

Element 1 is the Facility Designation.
Element 2 is the Facility Type (T1 or T3 etc.)
Element 3 lists the channel on which the sub-rate circuit rides.
Element 4 is the CLLI Code for the originating (called the "A" location) location.
Element 5 is the CLLI Code for the terminating (called the "Z" location) location.

A sample CABS CSR follows:

Figure 3.3

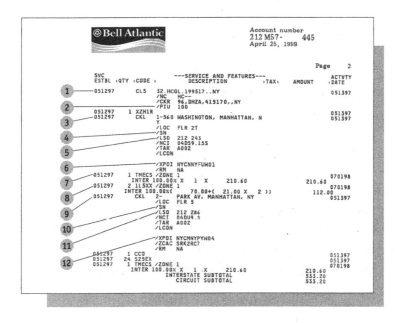

1   The USOC CLS tell us that the Circuit ID is in serial number format. The Circuit ID in this example is 32.HCGL.199517,,NY.

2   PIU 100 indicates that the service is used exclusively (100%) for interstate transmission.

3   CKL 1 provides the address of the originating location of the circuit.

4   The SN is the name of the customer at the originating end.

5   The Local Serving Office (LSO) is 212 243.

6   Point of interconnection.

7   The Channel Termination (CT) charge for the originating (A) location of the T1.

8   The 1L of 1L5XX always indicates mileage charges will follow. Here, under QTY (Quantity), the 2 indicates two miles. Interstate corridor mileage is billed via a standard formula. A fixed minimum mileage cost of $70.00 is added to the variable mileage cost of $21.00. The variable cost is determined by multiplying the total number of miles (2) by $21.00. The total mileage cost in this case is $112.00. The mileage is calculated by plugging the LSOs of the originating and terminating locations into a pricer.

9   CKL 2 provides the address of the terminating (Z) location of the circuit.

10   The SN is the name of the customer at the terminating end.

11   The Local Serving Office (LSO) is 212 286.

12   Point of interconnection for the terminating end.

Though the RBOCs are prohibited from providing interLATA service, certain exceptions are allowed. The FCC has designated certain areas as "corridor service". Contiguous areas of New York City and Northern New Jersey form the New York/New Jersey corridor, Camden/Philadelphia form another special corridor. Warren, Michigan and Adrian, Michigan are served by Ameritech and GTE respectfully and they are another example of areas considered corridor service.

Both corridor service and billing for CABS services that require billing by more than one LEC utilize what is called meet point billing. A handoff from one LEC to another LEC is performed at a certain meet point. The billing is calculated by using this point to calculate the percent of the circuit each LEC provides. On the CSR the BIP (Billing Percentage) FID tells the CABS system the proper per cent to bill. The following is a sample corridor CABS circuit:

BAN 212 M50-0001 Corridor Bill From Bell Atlantic - NY

	QTY	CODE	DESCRIPTION	RATE
1				
2		CLS	HCGY,999179,,NJ	
3			/PIU 100/NC HC--	
4		CKL	1-34 EXCHANGE PL,	
5			JERSEY CITY NJ	
6			/LSO 201 567	
7			/SN ABC NEWS	
8	10	1L5XX	/BIP 18	
9			(($70.00 X 18%)	
10			+ ($21.00 X 10 X 18%))	$ 50.40
11		CKL	1-172 MADISON AVE,	
12			MANHATTAN, NY	
13			/LSO 212 245	
14			/SN ABC CONTEMPORARY.	
15	1	TMECS	CHANNEL TERM	$210.60
TOTAL				$261.00

A corridor circuit will result in a separate matching bill from the other LEC with a different BAN but the same Circuit ID.

BAN 201 R16-0001 Corridor Bill From Bell Atlantic - NJ

1	QTY	CODE	DESCRIPTION	RATE
2	CLS		HCGY,999179,,NJ	
3			/PIU 100/NC HC--	
4	CKL		1-34 EXCHANGE PL,	
5			JERSEY CITY NJ	
6			/LSO 201 567	
7	1	TMECS	CHANNEL TERM	$210.00
8			/SN ABC NEWS	
9	10	1L5LS	/BIP 82 $60.00 + (10 X $17.70) X	
			82 BIP	$194.34
10		CKL	1-172 MADISON AVE,	
11			MANHATTAN, NY	
12			/LSO 212 664	
13			/SN ABC CONTEMPORARY.	
TOTAL				$404.34

When you are checking the billing for corridor circuits you need to make sure the BIPs add up to 100%. Each LEC will bill one Channel Termination. Each LEC will also bill Interoffice Mileage at their tariff rates times the BIP. You must also make sure each LEC comes up with the same Interoffice Mileage. In the example above both LECs calculate the Interoffice Mileage at 10 miles and the BIP from Bell Atlantic – NY (18%) plus the BIP from Bell Atlantic – NJ (82%) adds up to 100%.

If you have access to pricer software such as LATTIS you can independently check the rates. Simply plug in the LSOs of both locations, select the type of service (T1) and hit the enter key.

Figure 3.4

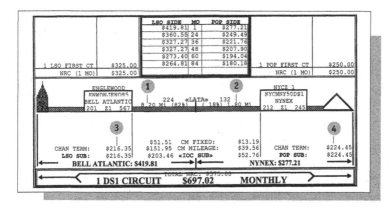

LATTIS will independently price out CABS circuits as well as special service circuits billed under state tariffs. Figure 3.4 confirms the following:
1 & 2. Interoffice Mileage on NJ side of the meet point is 8.20 miles and 1.80 miles on the NY side for a total of 10 miles. This matches the billing on the CSR.
3 & 4. LATTIS calculates different rates for the Channel Terminations and the Interoffice Mileage. LATTIS has a feature that allows you to access the appropriate tariffs.

Figure 3.5

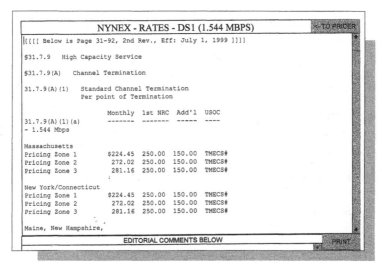

Figure 3.6

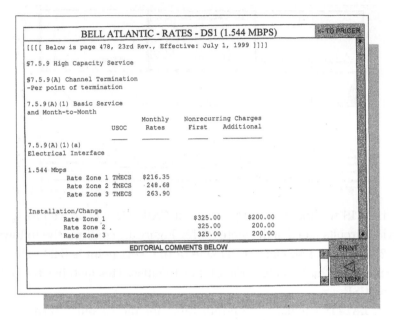

```
         BELL ATLANTIC - RATES - DS1 (1.544 MBPS)    ← TO PRICER
  [[[[ Below is page 478, 23rd Rev., Effective: July 1, 1999 ]]]]

  §7.5.9 High Capacity Service

  §7.5.9(A) Channel Termination
  -Per point of termination

  7.5.9(A)(1) Basic Service
  and Month-to-Month
                           Monthly   Nonrecurring Charges
                     USOC   Rates    First   Additional

  7.5.9(A)(1)(a)
  Electrical Interface

  1.544 Mbps
        Rate Zone 1 TMECS  $216.35
        Rate Zone 2 TMECS   248.68
        Rate Zone 3 TMECS   263.90

  Installation/Change
        Rate Zone 1                 $325.00   $200.00
        Rate Zone 2                  325.00    200.00
        Rate Zone 3                  325.00    200.00

              EDITORIAL COMMENTS BELOW                PRINT


                                                    TO MENU
```

Figures 3.5 and 3.6 are tariff excerpts from LATTIS. They reflect a rate change effective July 1st that was not reflected in the CSR. This should be reflected on next month's CSR.

Common CABS errors

The single most common error found on corridor service (and all interstate circuits) is the mis-application of the Special Access Surcharge Exemption (S25 USOC). FCC Tariff No. 40/41 Paragraph 5.7 (E)(2) allows telephone companies to place an additional surcharge on circuits that access the public switched network. This surcharge is automatically placed on your bill by the LEC unless you file an exemption.

It is your responsibility to inform the LEC if you are exempt from this surcharge. Your circuit is exempt from the special access surcharge if it can be characterized by any of the following:

- An open-end termination in a telephone company switch (FX line),

- An analog channel termination that is used for radio or television program transmission,

- A termination used for TELEX service,

- A termination, that by the nature of its operating characteristics could not make use of telephone company common lines,

- A termination, that interconnects either directly or indirectly to the local exchange network, where the usage is subject to Carrier Common Line charges, or,

- A termination, which the customer certifies to the LEC is not a PBX or other device capable of interconnecting the private line facility to a local subscriber line.

In essence, the circuit should be exempt if it is used for point to point transmission and can not access the public switched network. Since all but a small minorities of circuits are not exempt, most large companies and IXCs provide the LEC with a blanket exemption from the surcharge. When a blanket exemption is provided then it becomes the LEC's responsibility to make sure the surcharge is not applied.

On a CSR provided by Bell Atlantic - NY, you can identify the surcharge as follows:

CODE	DESCRIPTION	AMOUNT
S25	INTERSTATE 100%	$25.00

On a Bell Atlantic – NJ CSR, the surcharge is comprised of two charges, the S25 and the UTM USOCs.

CODE	DESCRIPTION	AMOUNT
S25	INTER NJ 100%	$25.00
UTM	INTER NJ 100%	$ 4.56

If you have an existing circuit and now realize it is surcharge exempt, notify the LEC. LEC representatives will issue three months back credit from the date they receive your exemption form. An exception is granted if you have proof that an exemption certificate was sent in at an earlier date, but was not applied to some of your accounts. Credit would be due back to the date of the blanket exemption letter up to a maximum of two years.

How to Identify and Obtain Refunds on
Corridor Circuits and Meet Point Billing

The key to auditing corridor service circuits is to compare the CSRs of both LECs providing corridor service. The rates may be different, but they should otherwise be mirror images of each other.

If for example, the Bell Atlantic - NY CSR does not have a S25 surcharge, but your Bell Atlantic - NJ CSR does have one, a bell should go off. You now have proof that Bell Atlantic - NY received an exemption form. To better understand why you are due credit, you need to know how these types of circuits are provisioned.

Step 1 - You call up Bell Atlantic - NY and request a voice grade data circuit between New York City and Northern New Jersey.

Step 2 - The Bell Atlantic – NY representative takes your request and enters your order into a software system called SAFE which generates what is known as an Access Service Request (ASR).

There are national standards that detail the format of an ASR. The ASR contains a field that lets all standard telephone company-billing systems know whether a Special Access Surcharge should be billed on a particular circuit. This field is called the S25 field. The representative must enter a Y (yes, charge the S25) or an N (no charge, the customer is exempt) in this field. The order is then sent to downstream Bell Atlantic – NY systems to provision the New York portion of the circuit. The order is simultaneously sent to Bell Atlantic – NJ via a mechanized data link.

Bell Atlantic – NJ downloads the order and converts it into a format its service order processor can read. If Bell Atlantic –NY has a Y instead of an N in this field, the Bell Atlantic – NJ order will also bill the S25 surcharge.

When negotiating refunds with either of the Bell companies, you can point out how complicated this system for coordinating orders can be. It also creates numerous opportunities for errors. You should not be penalized for errors caused by either telephone company.

Common Mileage Errors

The second most common corridor service error is the incorrect calculation of mileage. As previously mentioned, Bell Atlantic – NY and Bell Atlantic – NJ split mileage charges for a corridor circuit based on the portion of the circuit that lies in each state (the meet point).

The sum of the BIPs from Bell Atlantic – NY and Bell Atlantic – NJ should always equal 100%. The following example illustrates overbilling for mileage that is detected by adding BIPs on the Bell Atlantic – NY and Bell Atlantic – NJ CSRs. The Circuit ID is 32, LNGS,,999123. Compare the CSRs and add the BIPs together.

Bell Atlantic - NJ CSR

QTY	CODE	DESCRIPTION	AMOUNT
8	1L5XX	[$15.00 +(8 x.40) x 85BIP]	$17.72

Bell Atlantic - NY CSR

QTY	CODE	DESCRIPTION	AMOUNT
8	1L5XX	[$36.44+(8 x.4.24) x 45BIP]	$51.70

When you add up the BIPs (85 +45) they exceed 100%. In this example the customer is over billed and should get back credit from the date the circuit was installed. To calculate your savings first determine what the correct BIPs should be. Substitute the new BIPs into the mileage formula details on your CSR. In this example, Bell Atlantic – NY BIP should be 15% instead of 45%. Interstate circuit credit is limited to no more than two years per the Communications Act of 1934.

Other CABS Errors

Month to Month versus Term Plans

Most of the services provided by the LEC's FCC Access Tariffs are provided at a month to month standard rate or as a term plan which offers a discount if the customer commits to keeping the circuit for a certain time period. Since most IXCs have a need for large numbers of access circuits (DS1, DS3, etc.) a term plan makes sense over a month-to-month rate. A one-year term plan can yield a discount of 10% over the month-to-month rate, and increasing the commitment to 7 years can result in a discount of 35% over the month-to-month rate.

Common Term Plan Problems

Getting a copy of the signed Term Plan is half the battle. Many large companies loose or misfile the term plan contracts. LECs may also misplace contracts. Make sure all contracts signed with the LEC are kept in a safe, easy to access location for audit purposes.

Ameritech and Bell South have the capability to search their CABS database to identify all CABS BANs billed to a certain customer. That way you can be sure you are auditing all of your company's CABS accounts.

Single Payment Option (SPO)

The Single Payment Option discount plan is prone to serious billing errors. The SPO is similar to a Term Plan discount except that a customer makes a one-time payment up front for the entire term period. An additional discount is provided for this up front payment.

We have found many instances where the SPO was paid to the LEC but billing still continued. The LEC will want a copy of the canceled check that paid for the SPO in order to provide you with a refund.

Before you enter into standard term plans or pay a SPO be aware that termination charges apply if the service is terminated before the contract period is up. Termination charges vary by LEC and are set in their tariffs.

The following is an example of how Ameritech applies termination charges: A customer subscribing to a 60-month term plan disconnects their DS1 circuit during the 37th month. The customer is paying $250.00 per month based on a 60-month term plan agreement. The termination charge is calculated based on the difference between the 60-month term plan rate and a 36th month term plan rate as the 36th month term plan rate is the closest in length of time to the 37th month disconnect date.

DS1 circuit 60-month term plan rate is $250.00 per month; 36-month term plan rate is $276.80 per month. The calculation for termination charges is:

(36 month term plan rate – 60 month term plan rate) x 37 months
$276.80 - $250.00 x 37 months; or $26.80 x 37=$991.60

Since the DS1 circuit was disconnected in the 37th month the provider of the DS1 treated the DS1 circuit as if it was a 36-month term plan. In this example the customer is responsible for $991.60 in termination charges.

Chapter 4 provides additional examples of CABS billing errors.

Telecommunications Expense Management

How to Identify
Billing Errors and
Obtain Refunds

In this chapter we concentrate on identifying and correcting local telephone bill errors. This chapter will identify both CRIS and CABS system billing errors. Chapter 6 will show you how long distance errors are identified and long distance refunds are obtained.

Most LEC bill errors can be easily spotted by the careful review of your CSR. Checking the accuracy of your telephone bill does not have to be time intensive. The following guide to finding errors is both fast and accurate.

How to Identify Common Billing Errors

There is an easy way to find mistakes on your bills. Just follow these simple steps.

Step 1 – Obtain a list of all of your BTNs from your Accounts Payable Department. Large companies can also contact their LEC Account Executive to obtain a list of all the BTNs billed to your company. Bell Atlantic – NY refers to this list of BTNs as the MITAS list while Ameritech – Illinois refers to the same list as the AIMS list. Order both a CSR and duplicate bill for each BTN.

Step 2 - Write down all the telephone numbers that appear on each CSR.

Step 3 - Take special notice of any telephone line that has unique or different USOCs billed to it. For example, assume that your CSR lists billing for 37 POTS telephone lines. If 2 of the 37 lines have optional wire maintenance charges, a red flag should go up. As a general rule, you should check all deviations from the basic 1MB or ALN and the FCC Line Charge that typically make up the cost of a single POTS line. Repeat this process for your trunk lines.

Step 4 – Create a spreadsheet that lists all of your telephone numbers on each CSR. Call each telephone number. Your findings will generally fall into one or more of the following categories:

- Your call is answered. Make sure you verify that the person answering the phone works for your company. We have come across situations where a different company answers the call. In this case you must notify your LEC of the error and you would be due a credit for both the fixed monthly cost of the line and for all of the local and long distance calls made over that line. If the call is answered and the person works for your company note your spreadsheet that this line is OK.

- The telephone line is busy. Re-dial those telephone numbers that are busy at different times of the day. If still busy, note your spreadsheet that the line is busy.

- Your call is not answered. Redial telephone numbers that are not answered at different times of the day and on different days of the week. Check unanswered telephone lines against the line numbers listed on your long distance bill. If you use a call accounting system, check your old reports for usage on the lines in question. You can also call your LEC for help. Unlike long distance carriers, the LEC does not supply you with a list of your local calls with your bill. You can, however, call the Telephone Company and have them check your Local Usage Detail List (LUDs). Your LUDs itemize all your local calls, sorted by telephone line. Be careful how you word your request as there is an additional charge if you request a print of your LUD List (averages approximately $.50 per page). As an alternative to obtaining the LUDs, you can simply request the telephone company representative to visually check your LUD for usage on the telephone lines not recognized by you.

- The numbers that give off a tone when dialed are most likely connected to a FAX or modem line.

- You reach an intercept recording that advises the line is disconnected. This happens more frequently than many people realize. Many companies are paying for telephone lines that are not even in service. The obligation is on the LEC to find the date the service was actually disconnected. Note the results of your calls on your spreadsheet.

Step 5 - Diagram the originating and terminating locations of each special service circuit billed to your account. Check the terminating locations to determine if this location is still valid. For example, you may be paying for circuits terminating at a warehouse that moved cross-town or to a company you no longer do business with.

Step 6 – Use your PBX technician or in-house technician to check all of the lines and circuits that could not be verified by calling or by checking the numbers against your long distance calls and LUDs.

Here is the easy part. The telephone lines and circuits you have found to be unnecessary can be disconnected. Send a letter to your LEC representative authorizing their removal from your bill on a go forward basis. Obtain an order number for your records and follow-up to order a new CSR to make sure the billing has stopped for these lines or circuits.

Now the hard part. What if the telephone lines or circuits in question were disconnected three years ago? Check your records. Do you keep copies of disconnect letters sent to the LEC or a list of confirmation disconnect order numbers issued by the LEC? If your company has a dedicated LEC representative perhaps they remember that a location closed and all billing at that location should have stopped. Gather as much back-up material as you can.

Recently we audited a company that had moved out of a major location in January of 1995. While reviewing a bill, we identified 15 PBX tie lines connecting this location to corporate headquarters that were still in billing. The CSR showed that these PBX tie lines were connecting two PBXs and listed the main number at both locations. When I called the main number at the terminating location a completely different company answered it. The LEC had reassigned the main number. Even though we did not have the order number that should have disconnected the tie lines we were able to show that the customer had told the LEC it was shutting down its facilities at the terminating location. In this case the LEC agreed to provide full credit in the form of a refund check back to the date the main number was disconnected. The refund was over $100,000.00.

It's important to track the savings your company realizes on a go forward basis, as well as any refunds covered, as these savings contribute to the bottom line. Many times you will not have the paperwork to justify a refund. In that case your technician needs to check for dial tone or carrier signal at the demarcation point (the demarc). The demarc is defined as the point of interconnection of the LEC's communications facilities and terminal equipment, protective apparatus or wiring at a customer's location. In order for a telephone line to be considered operational there must be dial tone on the customer side of the demarc. See Figure 4.1.

Figure 4.1 Demarcation Block
(RJ21X Block)

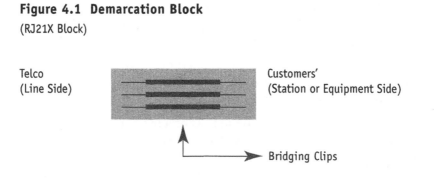

Telco   Customers'
(Line Side)   (Station or Equipment Side)

Bridging Clips

Your technician should verify that POTS lines, trunks and special service circuits appear on the LEC side of the demarc and are connected to the customer side via bridging clips. It is not uncommon for the LEC to physically remove your facilities but still leave the service in billing. LEC facilities are often reused for other services and sometimes for other customers. If you can prove that the LEC has physically removed or disconnected a telephone line or circuit then they have to show why or when the work was done. At that point you would be due out of service credit if you want the line re-installed or a refund if the line is not needed.

Step 7 - Send the list of telephone lines and circuits found to be non-operational at the demarc to the LEC. Also, provide the LEC with the list of any facilities billed by them that could not be found by your technician. The LEC has an obligation to "tag" (label and locate) the facilities they are billing you for and to prove they are operational. Insist they send out one of their technicians to your location. Even if the LEC's technician cannot locate some facilities or verifies that some billed facilities have been physically removed, the LEC will resist providing you with a refund. They will gladly stop the billing on a go forward basis and may offer three or six months credit just to placate you. If you are looking for additional credit you will have to do your homework. Start by researching the LEC's tariffs.

Tariffs

The Federal government regulates common carriers. A common carrier is a private company that sells or leases communications services and facilities to the public. Examples of common carriers include LEC's, long distance carriers and cellular providers. Common carriers must publish their rates and a description of their services with the government. Dominant carriers such as the RBOCs and AT&T are closely regulated to foster both universal service and an open, competitive market.

There are two levels of agencies. The Federal Communications Commission (FCC) regulates interstate (between states) and international communications. State Public Utilities Commissions (PUCs) regulate intra-state communications. This includes services with-in a LATA and services between LATAs within a particular state. Most of the services provided by the LEC's will be regulated by the PUC while the FCC will regulate most of the services provided by long distance carriers.

Tariffs contain thousands of pages and are constantly being changed and updated. Many State tariffs will provide information on overbilling claims and the LEC's liability. Figures 4.2, 4.21 and 4.22 are pages from Bell Atlantic-NY's (AKA New York LEC) state tariff known as the PSC

Figure 4.2

P.S.C. No. 900--Telephone

New York Telephone Company

Section 1
5th Revised Page 117.2
Superseding 4th Revised Page 117.2

General Rules And Regulations

H. Payments And Termination Of Service (Cont'd)

16. Billing Discrepancies

General

The following provisions govern the disposition of billing discrepancies related to recurring monthly charges for exchange access lines, private lines, service features, such as TOUCH-TONE Calling Service, and Custom Calling Service, and equipment. Except for the provision or interest (as set forth in subparagraph (3)), these provisions shall not apply to charges related to the Company's usage services, including access to interexchange carriers.

Overbilling

1. When a claim for overbilling is made to the Company and that claim is verified by a physical inventory or other Company verification procedure, credit is given to the date of disconnection, if available from Company records or reasonable evidence provisioned by the customer, up to a maximum of six years. If a disconnect record is not available, but a record of physical activity other than disconnection is available, credit is given to the date of that activity, up to a maximum of three years. Where neither a disconnect record nor a record of other activity is available, credit is given to the date billing commenced, up to a maximum of three years.

2. For purposes of the above tariff provision, an exchange access line and its associated features and equipment shall be deemed to still be in service, where dial tone exists at the customer's demarcation point, as defined in Section 16 of the Tariff. For purposes of the above tariff provision, a private line and its associated features and equipment shall be deemed to still be in service where a continuous dedicated circuit exists between the customer's demarcation points. (C)

Issued: February 19, 1996   Effective: May 4, 1996
By Sandra DiIorio Thom, General Attorney
1095 Avenue of the Americas, New York, N.Y. 10036

Figure 4.21

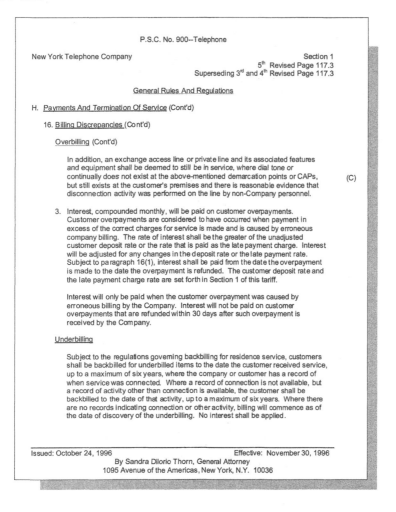

P.S.C. No. 900--Telephone

New York Telephone Company   Section 1
                                          5th Revised Page 117.3
                              Superseding 3rd and 4th Revised Page 117.3

General Rules And Regulations

H. Payments And Termination Of Service (Cont'd)

16. Billing Discrepancies (Cont'd)

Overbilling (Cont'd)

In addition, an exchange access line or private line and its associated features
and equipment shall be deemed to still be in service, where dial tone or
continually does not exist at the above-mentioned demarcation points or CAPs,   (C)
but still exists at the customer's premises and there is reasonable evidence that
disconnection activity was performed on the line by non-Company personnel.

3. Interest, compounded monthly, will be paid on customer overpayments.
Customer overpayments are considered to have occurred when payment in
excess of the correct charges for service is made and is caused by erroneous
company billing. The rate of interest shall be the greater of the unadjusted
customer deposit rate or the rate that is paid as the late payment charge. Interest
will be adjusted for any changes in the deposit rate or the late payment rate.
Subject to paragraph 16(1), interest shall be paid from the date the overpayment
is made to the date the overpayment is refunded. The customer deposit rate and
the late payment charge rate are set forth in Section 1 of this tariff.

Interest will only be paid when the customer overpayment was caused by
erroneous billing by the Company. Interest will not be paid on customer
overpayments that are refunded within 30 days after such overpayment is
received by the Company.

Underbilling

Subject to the regulations governing backbilling for residence service, customers
shall be backbilled for underbilled items to the date the customer received service,
up to a maximum of six years, where the company or customer has a record of
when service was connected. Where a record of connection is not available, but
a record of activity other than connection is available, the customer shall be
backbilled to the date of that activity, up to a maximum of six years. Where there
are no records indicating connection or other activity, billing will commence as of
the date of discovery of the underbilling. No interest shall be applied.

Issued: October 24, 1996   Effective: November 30, 1996
            By Sandra DiIorio Thorn, General Attorney
            1095 Avenue of the Americas, New York, N.Y. 10036

900 tariff. These tariff pages spell out the requirements that prove an
overbilling situation and the time period for which credit will be calculated.

A careful reading of these tariff pages show that with proof refunds
will be backdated up to six years. If a LEC does not spell out the time
period for refunds in it's tarrif's it is usually equal to the statue of limita-
tions in that state. If you do not have proof but can show that there is no

dial tone or carrier signal at the demarc then three years credit will be given.

This tariff also spells out the interest due the customer for the money collected in error by the Bell Atlantic-NY. It is 18%. Where a LEC does not spell out the interest it will pay on overbilling claims in its tariffs we request that the interest equal the late payment interest rate charged by the LEC. Other company's tariffs have different regulations.

Figure 4.22

P.S.C. No. 900--Telephone

New York Telephone Company

Section 16
5th Revised Page 7.1
Superseding 3rd and 4th Revised Pages 7.1

Explanation Of Terms

Demarcation Point

The point of interconnection of the Telephone Company communications facilities and terminal equipment, proactive apparatus or wiring at a customer's location. (C)

- Building Owner Demarcation Point Determination

 A building owner may determine the demarcation point for new wire or modification or addition to existing wire provided, however, that the demarcation point cannot be further than twelve (12) inches from where the telephone wiring enters the customer's premises.

- Residence Customers

 The demarcation point for a single unit residence customer is located within twelve (12) inches of the protector, or if there is no protector, within twelve (12) inches of where Telephone Company wiring enters the customer's premises. An existing SNI/NI shown on Company records will not be relocated to the demarcation point with the effective date of this Tariff except as specified in Section 1, Paragraph B.1.1. b. (2) of this Tariff.

 The demarcation point for a multi-unit residential customer is located within the customer's premises at a point no further than twelve (12) inches from where the Telephone Company wiring enters the customer's premises. An existing SNI/NI shown on Company records will not be relocated to the demarcation point with the effective date of the Tariff except as specified in Section 1, Paragraph B. 1. 1. b. (2) of this Tariff.

- One and Two Line Business Customers

 The demarcation point for a single unit business customer subscribing to one or two lines located within twelve (12) inches of the protector, or if there is no protector, within twelve (12) inches of where Telephone Company wiring enters the customer's premises. An existing SNI/NI shown on Company records will not be relocated to the Demarcation Point with the effective date of the Tariff except as specified in Section 1, Paragraph B1. 1. b. (2) of this Tariff.

 The demarcation point for a multi-unit business customer subscribing to one or two lines is located within the customer's premises at a point no further than twelve (12) inches from where the Telephone Company wiring enters the customer's premises. An existing SNI/NI shown on Company records will not be relocated to the demarcation point with the effective date of this Tariff except as specified in Section 1, Paragraph B. 1. 1. b. (2) of this Tariff.

- Multiline (more than two lines) Business and Private Line Business Customers in Single Tenant Buildings

 • Buildings with New York Telephone Riser Cable

 The demarcation point is located within twelve (12) inches of each existing House/Riser Terminal.

 • Buildings without New York Telephone Riser Cable

 The demarcation point is located within twelve (12) inches of the protector or, if there is no protector, within twelve (12) inches of where the Telephone Company's wire enters the building.

Issued: October 24, 1996

Effective: November 30, 1996

By Sandra DiIorio Thorn, General Attorney
1095 Avenue of the Americas, New York, N.Y. 10036

Figure 4.3 is a page from Ameritech's Michigan tariff that itemizes a two-year refund period. This change was recently made in response to an audit that was performed by us on Ameritech's largest customer and is clearly designed to limit Ameritech's liability.

Figure 4.3

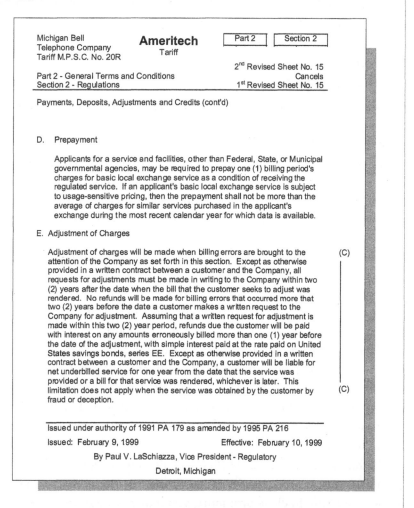

Michigan Bell Telephone Company Tariff M.P.S.C. No. 20R

Ameritech Tariff

Part 2   Section 2

2nd Revised Sheet No. 15

Part 2 - General Terms and Conditions
Section 2 - Regulations

Cancels
1st Revised Sheet No. 15

Payments, Deposits, Adjustments and Credits (cont'd)

D. Prepayment

Applicants for a service and facilities, other than Federal, State, or Municipal governmental agencies, may be required to prepay one (1) billing period's charges for basic local exchange service as a condition of receiving the regulated service. If an applicant's basic local exchange service is subject to usage-sensitive pricing, then the prepayment shall not be more than the average of charges for similar services purchased in the applicant's exchange during the most recent calendar year for which data is available.

E. Adjustment of Charges

Adjustment of charges will be made when billing errors are brought to the attention of the Company as set forth in this section. Except as otherwise provided in a written contract between a customer and the Company, all requests for adjustments must be made in writing to the Company within two (2) years after the date when the bill that the customer seeks to adjust was rendered. No refunds will be made for billing errors that occurred more that two (2) years before the date a customer makes a written request to the Company for adjustment. Assuming that a written request for adjustment is made within this two (2) year period, refunds due the customer will be paid with interest on any amounts erroneously billed more than one (1) year before the date of the adjustment, with simple interest paid at the rate paid on United States savings bonds, series EE. Except as otherwise provided in a written contract between a customer and the Company, a customer will be liable for net underbilled service for one year from the date that the service was provided or a bill for that service was rendered, whichever is later. This limitation does not apply when the service was obtained by the customer by fraud or deception.

(C)

(C)

Issued under authority of 1991 PA 179 as amended by 1995 PA 216

Issued: February 9, 1999   Effective: February 10, 1999

By Paul V. LaSchiazza, Vice President - Regulatory

Detroit, Michigan

Even if a tariff limits a customer's liability that does not mean a LEC can't provide credit beyond this limitation. When they are working with large clients the LEC recognizes that these companies have more choices than in the past. The LEC also realizes that their billing systems are over thirty years old and are error-prone. Additionally, if a contract is signed that specifies certain rates, the LEC also has a legal obligation to honor its commitments.

Getting back to obtaining a refund, you now know that three months credit is unacceptable. You can send a copy of the tariff pages that deal with overbilling with your billing claim to obtain the maximum refund possible.

Tariff Errors

Billing errors fall into three major categories. The first major category is disconnected lines or circuits that still appear on your bill, the second category is rate and tariff errors while the third category is contract errors.

Special service circuits often have tariff errors. Most companies have special rates that should apply when circuits connect buildings located within a campus-like environment, are contiguous or originate and terminate in adjacent city blocks. Special USOCs need to be entered when these situations arise.

To identify these kinds of errors, you need to be able to understand the general rules for special service billing and then verify the specific charges against what is in a LEC's tariffs. Once you get accustomed to a LEC's billing practice in a certain area, you will be able to verify charges simply with a quick review of the charges listed on the CSR.

The standard billing for special service circuits includes two local channels and interoffice mileage (if applicable). To start your verification, you need to look at the LSOs on CKL 1 and CKL 2 to determine if the circuit in question originates and terminates within the same CO (intra-office circuit). If two different COs serve the circuits then interoffice mileage applies.

If the circuit is an intra-office circuit, inter-office mileage charges should not apply. Additionally, if the originating and terminating points of the circuit are in close proximity, you may not be responsible for two local channel charges (i.e. CON2X or CON4X-type USOCs in Bell Atlantic-NY). Figures 4.4 and 4.5 illustrate the differences between a Central Office Loop and a Block Loop.

Figure 4.4 Central Office Loop Charge

(CON2X) applies to the line between the pole or terminal box and the local Central Office Building.

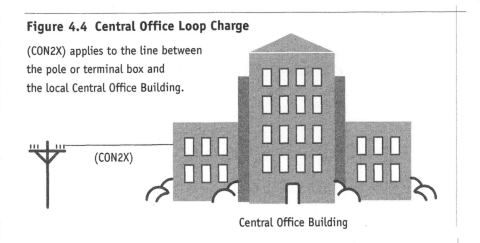

Central Office Building

Figure 4.5 Block Loop Charge

(BKP2X) applies to the line between different customer locations on the same or adjacent city blocks.

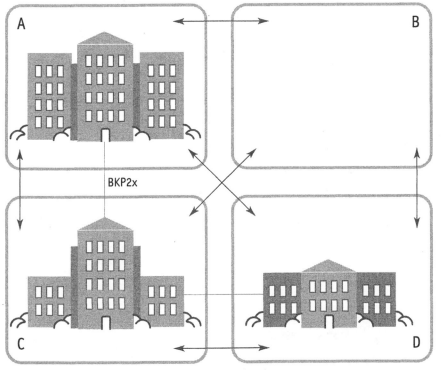

Note: Blocks A, B, C, and D are adjacent to each other.

Figure 4.5a Feature Function Charge

(PMWV2) applies to the line from the pole or terminal box to the interface on the customer premises.

Customer Premises

Example #1 - In this example the OSNA circuit both originates and terminates within the same central office area. Both locations are served by the same CO and are not located on adjacent city blocks. Since the same CO serves both locations, monthly interoffice mileage charges are not applicable. Figure 4.6 illustrates how this circuit is designed.

Figure 4.6 Billing for an OSNA Circuit Located in Same CO

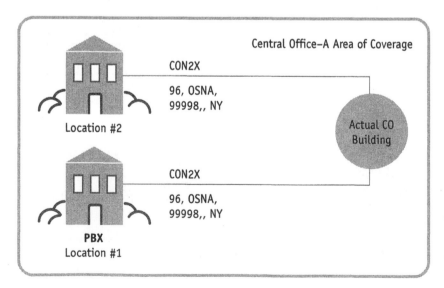

QTY	ITEM	MONTHLY
2	CON2X	$43.06

Example #2 - The monthly cost for an OSNA circuit that is within the same CO, and is also located on an adjacent block is charged one BKP2X instead of two CON2Xs.

Figure 4.7 OSNA Circuit-
Adjacent Blocks

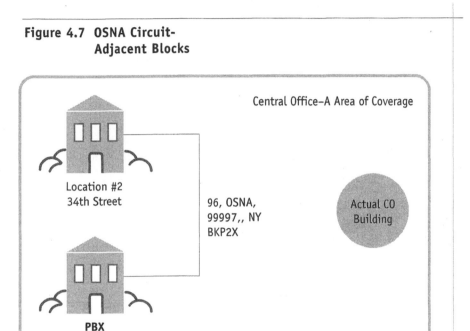

Central Office–A Area of Coverage

Location #2
34th Street

96, OSNA,
99997,, NY
BKP2X

Actual CO
Building

PBX
Location #1
35th Street

QTY	ITEM	MONTHLY
1	BKP2X	$27.73

Each local LEC has its own billing nuances. Bell Atlantic-NJ has different local channel rates dependant on whether the circuit is designed as an intra or inter-office circuit. The following details differences in intra and inter-office billing as they appear on Bell Atlantic-NJ CSRs.

Example #3 - Intra-office PXOS (PBX outside extension) circuit. Monthly charges as billed by New Jersey Bell. This example details the billing as it appears on a Bell Atlantic – NJ CSR. Figure 4.8 details how this circuit is designed.

**Figure 4.8 Bell Atlantic NJ
            Intra Office OSNA Circuit**

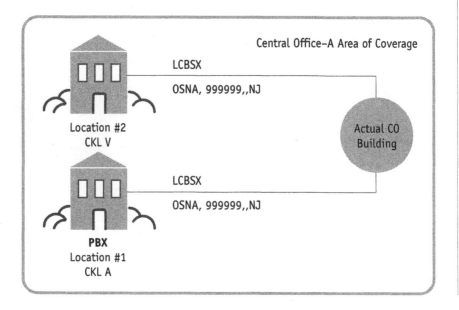

USOC	DESCRIPTION	MONTHLY
CKT	OSNA 999999NJ	
ARR	PX 3XAYS/CKL A, V	
	SAY /**TYPE C SIGNALING ARRANGEMNT	$.15
	LCBSX/CKLA/LSO 609 999/**LOC CHAN SM	
	EXCH	$21.73
	LCBSX/CKL V/LSO 609 555/**LOC CHAN SM	
	EXCH	$21.73
	TOTAL	$43.61

Example #4 - If the OSNA-type circuit connects locations served by different COs (inter-office) Bell Atlantic – NJ will bill as follows (COs are six miles apart in this example):

Figure 4.9 Bell Atlantic NJ
              Inter Office OSNA Circuit

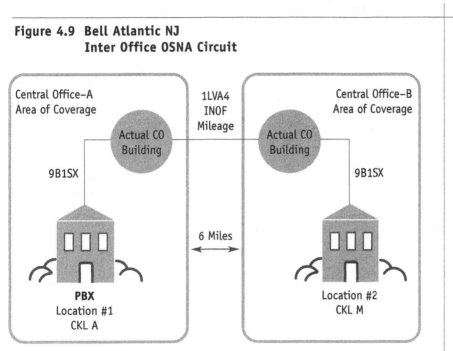

USOC	DESCRIPTION	MONTHLY
CKT	OSNA 999998NJ	
ARR	PX 4XAYS/CKL A, M	
	SAY/**TYPE C SIGNALING	
	ARRANAGEMENT	$.15
	9B1SX/CKL A/LSO 609 999/**	
	LOC CHAN DIF EXCH	$23.46
	9B1SX/CKL M/LSO 609 999/**	
	LOC CHAN DIF EXCH	$23.46
6	1LVA4/SEC 1/EX WAYNE	
	MILEAGE INTEREXCHANGE	$28.42
	TOTAL	$75.49

The monthly cost for this circuit totals $75.49. Figure 4.9 illustrates how this circuit is designed.

Example #5 - If you connect 2 PBXs (via a tie line) located in the same CO, Bell Atlantic – NJ will bill you as follows:

Figure 4.10 Bell Atlantic NJ
Tie Line Intra Office

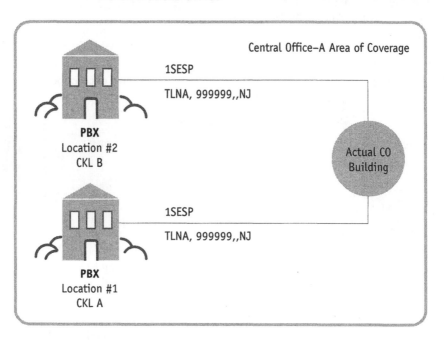

USOC	DESCRIPTION	MONTHLY
CKT	TLNA 999999NJ	
ARR	PX 3XAYS/CKL A,B	
	SLM/CKL A/**E&M TYPE	$ .90
	1SESP/CKL A/LSO 609 999/**	
	LOC CHAN SM EXCH	$27.64
	SLM/CKL A/**E&M TYPE	$ .90
	1SESP/CKL A/LSO 609 999/**	
	LOC CHAN SM EXCH	$27.64
	TOTAL	57.08

The monthly cost of this circuit totals $57.08. Figure 4.10 illustrates how this circuit is designed.

Example #6 - If you connect 2 PBXs located in different COs, Bell Atlantic – NJ will bill you as follows:

Figure 4.11 Inter Office Tie Line

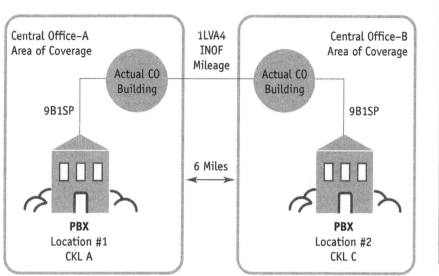

USOC	DESCRIPTION	MONTHLY
CKT	TLNC999998NJ	
ARR	PX 3XAYS/CKL A,C	
	SLM/CKL A/**E&M TYPE	$ .90
	9B1SP/CKL A/LSO 609 999/**	
	LOC CHAN DIF EXCH	$29.08
	SLM/CKL A/**E&M TYPE	$ .90
	9B1SP/CKL C/LSO 609 999/**	
	LOC CHAN DIF EXCH	$29.08
6	1LTA4/SEC 1/EX WAYNE	
	INTEROFFICE MILEAGE	$28.42
	TOTAL	$88.38

The monthly cost of this circuit totals $88.38. Figure 4.11 illustrates how this circuit is designed.

Sample CSR & Bill Errors

The following are sample claims submitted by professional auditors. These claims cover a wide variety of local telephone billing errors from a variety of vendors. In later chapters we will provide sample errors found in Carrier Access Bill Systems (CABS) and in long distance vendor bills.

Case Study # 1

An audit of Centrex lines confirmed that a Michigan based client was overbilled:

A bank in Michigan had 130 Centrex lines. Ameritech – Michigan's Centrex tariffed rates were on a five-tier scale as follows:

1st 25 lines @	$10.03
Next 174 lines @	6.32
next 300 lines @	5.49
next 500 lines @	4.57
add'l lines @	3.59

A review of the CSR found that all 130 Centrex lines were billed at the $10.03 rate per line for a period of just over six years. After an audit and confirmation of the billing error, this client received a refund of $28,000.00.

Figure 4.12a

CUSTOMER SERVICE RECORD

Bell Atlantic

				ACCOUNT NUMBER				0 4 8

				CLASS OF SERVICE	DIRECTORY	PAGE	
				E 7 K J X	W		

BILL PERIOD		RECORD SEGMENT			PRINT DATE		PRINT REA
JUNE 1 - JUNE 30, 1999		ACCOUNT			6-3-99		B

QUANTITY	SERVICE	DESCRIPTION		L	ACTIVITY DATE	TOTAL	TAX	A
	PCL	LOCL						
	OTN							
	ZBCS	B3-CBV						
	LN			I	8-27-98			
	LA							
	SA			I	1-31-96			
				I	1-31-96			
	LOC	DES STORE		I	1-31-96			
	YPH	EA495 ELECTRONICS/SIC 5064		I	1-31-96			
		---BILL						
	BN1				12-11-98			
	BN2	INVESTMENTS INC			12-11-98			
	BA1				12-11-98			
	PO							
		08817			12-11-98			
	LB	02108						
	BILP	01						
	CSG	MB3,B						
	TAR	068						
		---RMKS						
	RMKR	1-147		I	1-31-96			
		---S&E						
1	E7KJX			I	1-31-96			
1	NYKNP	/SPP CSP						
		/TA 60 MO, 09-29-94						
		/SBA 940153/CN 01						
		/ZLSZ 00046/NOL 01100/OV						
		(IntelliPath - Digital)		I	1-31-96	509.22	4	
1	PGJB1	(FCC Line Charge Offset)		I	7-20-98		1	
1	SBSBX	(Summary Billing)		I	9-20-96		2	
1	SBSBX	(Summary Billing)		I	9-24-96		2	
1	VWDX3	/RTE 0.00/SPP CTA						
		/TA 60 MO, 09-03-96						
		(Large Volume Discount 17%)		I	9-19-96		1	
1	JJS5S	(Single-Line Jack)		I	7-11-96		4	
1	RJ21X	(25-Line Connector Jack)		I	1-31-96		4	
1	RJ21X	(25-Line Connector Jack)		I	1-31-96		4	
		(CONT')						

RECORD SEGMENT	ACCOUNT		CSG MB3	ACCOUNT NUMBER	

Case Study # 2

An audit of Centrex lines confirmed that a New York based client was entitled to a $450,000.00 refund.

In auditing a Centrex account in Bell Atlantic – NY the following errors were identified by careful review of the CSR (see figures 4.12a & b):
1. The /ZLSZ FID tells us this particular location has 46 lines. The /NOL FID (Number of Lines) tells us that this location is part of a larger Centrex contract for 1100 Centrex lines.

Figure 4.12b

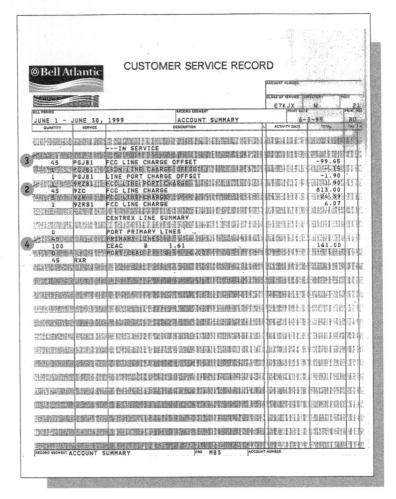

2. The 45 9ZC charges total $813.00. By dividing the total by the number of lines, we find that there is a charge of $18.06 per 9ZC. The tariff shows that the 9ZC should be $8.13.

3. A credit is given in the amount of $99.45 to reduce the 9ZC charge. The 9ZC charge is still $15.85 instead of $8.13.

4. Customer is billed 100 CEAC charges instead of 45. 55 lines were billed at $1.61 in error.

The other BTNs under the contract were identified and obtained and they all had exactly the same problem. A claim was issued on all 1100 lines and a refund of over $500,000.00 was obtained for this customer.

Case Study # 3

An audit of special service circuits confirmed that a Connecticut based client was entitled to a $400,000.00 refund.

A large insurance company was found to be billed for two local channel charges on approximately 100 low speed voice and data circuits between buildings at its campus like headquarters. A review of the SNET tariffs found that these circuits should have been billed one intra-building charge. Our technician had traced these circuits out and found that they directly connected each of the buildings via direct cable runs.

Case Study # 4

An audit of special service circuits confirmed that a New York based client was entitled to a $600,000.00 refund. See figures 4.13a-4.13c.

A large hospital was billed two local channel charges on approximately 150 low speed voice and data special service circuits between a number of buildings that were across the street from each other. A schematic of the hospital's buildings found three tunnels under the street owned by the hospital that directly connected all of the buildings. A review of the Bell Atlantic – NY tariffs found that these circuits should have been billed cable carrying charges. Our technician had traced these circuits out and found that they directly connected each of the buildings via direct cable runs.

Case Study # 5

Bell Atlantic – NJ customer billed incorrect T1 handoff rate resulting in $21,000.00 refund

An Information Services company had three switched T1s to its CO dating back to 9/96. The billing for a switched T1 in most LECs has a charge for the T1 facility and either a port charge into the LEC's digital switch (DMS 100 or 5 ESS) or what is called a T1 handoff into the digital

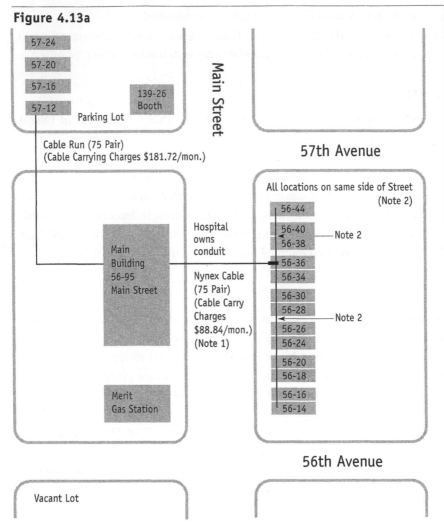

Figure 4.13a

57-24

57-20

57-16

57-12

139-26
Booth

Parking Lot

Main Street

Cable Run (75 Pair)
(Cable Carrying Charges $181.72/mon.)

57th Avenue

All locations on same side of Street
(Note 2)

56-44

56-40
56-38 — Note 2

Main
Building
56-95
Main Street

Hospital
owns
conduit

56-36

56-34

Nynex Cable
(75 Pair)
(Cable Carry
Charges
$88.84/mon.)
(Note 1)

56-30

56-28 — Note 2

56-26

56-24

56-20

56-18

Merit
Gas Station

56-16

56-14

56th Avenue

Vacant Lot

Notes:

1. Hospital owns conduit that runs between 56-95 Main Street and 56-36 Main Street.
    Nynex owns cable and is charging Cable Carrying Charges only from 56-95
    Main Street to 56-36 Main Street.

2. For Locations at 56-14, 56-24, 56-26, and 56-44 Main Street Bell Atlantic-NY is
    Charging BKP2X (Block Loop Charges). Nynex 900 tariff section 7, E.1 (page 52)
    and and E.4.c.1 (page 55) states that the customer has the option to pay Block
    Loop Charge or Cable Carrying Charges (E.1) when local facilities are used as
    defined in (E.4.c.1) as direct wire runs or local cable pairs on Telephone Company
    poles within the block and along either side of the adjoining streets.

Figure 4.13b

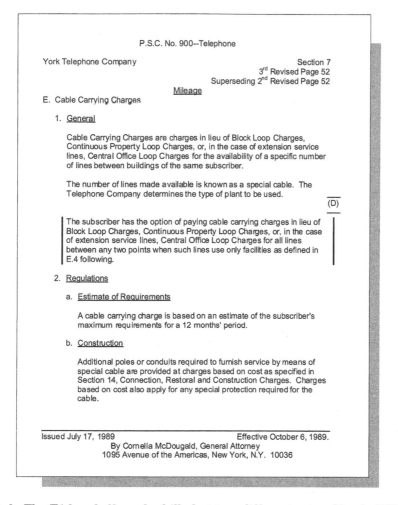

P.S.C. No. 900--Telephone

York Telephone Company

Section 7
3rd Revised Page 52
Superseding 2nd Revised Page 52

Mileage

E. Cable Carrying Charges

1. General

Cable Carrying Charges are charges in lieu of Block Loop Charges, Continuous Property Loop Charges, or, in the case of extension service lines, Central Office Loop Charges for the availability of a specific number of lines between buildings of the same subscriber.

The number of lines made available is known as a special cable. The Telephone Company determines the type of plant to be used.

(D)

The subscriber has the option of paying cable carrying charges in lieu of Block Loop Charges, Continuous Property Loop Charges, or, in the case of extension service lines, Central Office Loop Charges for all lines between any two points when such lines use only facilities as defined in E.4 following.

2. Regulations

a. Estimate of Requirements

A cable carrying charge is based on an estimate of the subscriber's maximum requirements for a 12 months' period.

b. Construction

Additional poles or conduits required to furnish service by means of special cable are provided at charges based on cost as specified in Section 14, Connection, Restoral and Construction Charges. Charges based on cost also apply for any special protection required for the cable.

Issued July 17, 1989   Effective October 6, 1989.
By Cornelia McDougald, General Attorney
1095 Avenue of the Americas, New York, N.Y. 10036

switch. The T1 handoff can be billed at two different rates. If only DIDs are transported on the switched T1 the USOC is D7Z and the monthly rate is $91.00. If two way trunks or a combination of DID and combo trunks are transported on the T1 then the correct USOC is D7W and the monthly rate is $292.00. In this case the customer was billed $292.00 for each T1 handoff even though the T1s transported only DID trunks. The refund was calculated for the difference between $292.00 and $91.00 for each handoff back to 9/96.

Figure 4.13c

P.S.C. No. 900--Telephone

New York Telephone

Section 7
Original Page 55

Mileage

E. Cable Carrying Charges (Cont'd)

  4. Definitions

    Local Facilities:

    a. Local pairs in block cable.

    b. Direct wire runs or local cable pairs on the subscriber's poles, or in the subscriber's conduits or in passageways when such poles, conduits or passageways are on the subscriber's property or rights-of-way including street crossings for which the subscriber has a permit.

    c. Direct wire runs or local cable pairs on Telephone Company poles within the following limits:

      (1) Within the block and along either side of the adjoining streets.
      (2) Between two adjacent blocks including blocks diagonally across street intersections.
      (3) Where blocks are not established, 1/10 mile or less of general distributing poles with any number of other poles.

Issued May 3, 1982.

Effective June 4, 1982.

By General M. Oscar, General Attorney
1095 Avenue of the Americas, New York, N.Y. 10036

Case Study # 6

Pacific Bell – California based customer billed incorrect T1 contract rates resulting in $500,000.00 refund.

This customer had negotiated a T1 flat rate of $233.00 per T1 provided they maintained over 100 T1s with Pacific Bell. Pacific Bell billed all of these T1s under one BTN. Subsequently, the Customer disconnected 15 of the T1s billed to that BTN. They had added 21 T1s that were billed to another BTN. The Pacific Bell billing system reverted the 85 T1s to the standard tariff rate as it did not take into account the customer still maintained over 100 T1s total with Pacific Bell. The average tariff rate was approximately $425.00.

Case Study # 7

Ameritech- Illinois based customer billed 9ZR charges on point to point T1s resulting in $20,000.00 refund.

Point to point T1s should not be billed 9ZRs because 9ZRs only apply to access lines. This customer had a combination of switched T1s (which should be billed 9ZRs) and point to point T1s which should not be billed 9ZRs.

Case Study # 8

GTE – Texas based customer billed voice mail charges in error on 20,000 Centrex lines resulting in a $2,840,000.00 refund.

An audit of a special custom contract found that voice mail should have been included in the Centrex line rate and was being billed separately. The refund went back to 1993 (the date the contract was signed).

Figure 4.14

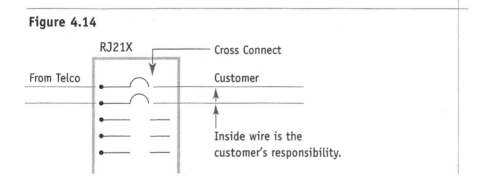

Case Study # 9

BellSouth – Georgia refunds for wire maintenance. See figure 4.14.

In BellSouth, it is possible to claim the inside wire maintenance USOC when it is billed on the CSR. The claim should state that the line/trunk terminates on a RJ21X demarc and as such, the inside wire which is located on the customer's side of the demarc is the customer's responsibility. Inside wire is maintained by the customer's vendor and not BellSouth.

Note that the inside wire maintenance charge is valid if the telco line terminates on a RJ11 or RJ45 so you want to specifically state that it terminates on a RJ21X.

Case Study # 10

Bell Atlantic – NY "dangling circuit" results in a refund for $4,000.00.

QTY	ITEM	DESCRIPTION	ACT DTE	RATE
1	PFSAS	/CKT 96,BANA,,99916,,NY	09-22-87	
	CKL	1-444 BROADWAY, MANHATTAN, NY/DES FLR 3/LSO 212 966/CLS 96,BANA,,99916,,NY		
1	CON2X	/EAR (LOOP CHARGE 2-WIRE)	2-1-90	$21.53
1	NRASQ	/DES 42A BLK HOLMES INSTR SOUTH STAIRWELL		
1	PMWV2	/EAR (BASIC 2-WIRE VOICE CIRCUIT)	2-1-90	$ 7.07

An examination of this CSR reveals that only one end of the circuit is listed (CKL 1). All circuits must connect at least two locations. CKL 1 is what is left of a partial disconnect. When the LEC disconnected the circuit they only stopped the billing of the terminating end (CKL 2). Billing for CKL-1 continued. The customer is due a refund back to the date of the partial disconnect order.

Case Study # 11

Bell Atlantic – NJ Remote Call Forwarding (RCF) charges left in billing results in $10,000.00 refund.

If your company has moved in the last five years, check your CSR for RCA or RCF numbers. RCA and RCF are the USOCs associated with Remote Call Forwarding. Remote Call Forwarding automatically forwards calls from the dialed number to another designated telephone number. When companies move, they often keep some of their old telephone numbers to forward calls to their new telephone number for a certain amount of time. This service is similar to call forwarding except that it is programmed at the CO and stays in effect all the time.

This service (and its associated billing) is usually designed to end at a prearranged date. Unless you check your CSR, billing for this service can continue for years. A CSR review for one of our clients located in New Jersey found the following (partial extract of the full CSR):

ACT DATE	USOC	QUANT	DESCRIPTION	RATE
06-17-87	RCA	5	/TN 201 999-2990/LSO	
			908 555/FSO	
			908 999-1010/	
			REMOTE CALL	
			FWDING	$72.60

The CSR extract above lists a TN of 201 999-2990, yet our client was located in the 908 area code. We dialed 201 999-2990 which forwards the call to 908-999-1010 and received a disconnect recording. We found that 908 999-1010 had been a temporary number set up at the new site until the PBX was up and running. Further investigation revealed that the customer had moved to their current location on 06-17-87 and that 201 999-2990 was their telephone number at the old location. The quantity (5) listed on the CSR meant that up to five simultaneous telephone calls could be forwarded from the old telephone number to the new one. Once we alerted the LEC to this error, the customer was credited $72.60 (plus taxes and surcharges) per month back to 12-17-87.

Tax and Surcharge Claims

Before we provide refund examples on taxes and surcharges. Some background information is helpful.

Taxes

Tax errors include state, local and federal excise taxes (a tariff of 3% on telecommunications services). For example, common carriers, telephone and telegraph companies and radio and television broadcasting stations are exempt from federal excise tax under Section 4253 (f) of the Internal Revenue Tax Code on their WATS service. Many trucking and shipping companies qualify as common carriers and are paying this tax in error.

Other exemptions to the federal excise tax include:
• tax exempt government organizations,
• nonprofit educational organizations,
• schools operated by churches or other religious bodies,
• nonprofit hospitals,
• consulate offices of foreign governments, and,
• local organizations conducting the community action portion of an economic opportunity program.

Dedicated private network services are also exempt from this tax. Exemptions from state and local taxes vary by state, but federal tax exemptions apply across all states. Tax billing is driven by the USOCs that appear in the BILL Section of your CSR. For example, a Bell Atlantic – NY customer that is federal, state and local tax exempt will have a BILL section that appears as follows:

ITEM	DESCRIPTION
BN1	RC HOSPITAL
BN2	HEALTH
BA1	781 E 42 ST
PO	NY NY 10036
TAR	002
TAX	FEDTELOCTE

The notation "FEDTELOCTE" after the USOC TAX tells us this customer is federal, state and local tax exempt. This tax-exempt indicator should also adjust billing of taxes on long distance calls that appear on your local LEC bill. You should check this and also check any long distance bills that are separately provided by a long distance carrier.

Tax refund policies vary by LEC. Bell Atlantic - NY, for example, will refund up to three years of federal taxes, but only three months of state and local sales tax paid in error. Bell Atlantic – NJ will not refund any back taxes to you, while AT&T will refund taxes paid over the last two years. These companies will require you to send them proof of your exemption before they will refund taxes and stop future tax billing.

In any case, you can always obtain taxes paid in error directly from the federal government by filing IRS form 843. The IRS will refund taxes paid in error over the last three years. They require a log detailing the telephone taxes paid, and you must make your old telephone bills accessible in case your refund request is audited.

You can recover state and local taxes paid in error from the appropriate state agency. In New Jersey, for example, you can file form A-3730 to obtain a tax refund. State forms and refund policies vary state by state, but three years back credit is most common. When you discover the incorrect billing of taxes, make sure you notify your LEC to prevent the future billing of taxes. Follow-up and check all future bills.

Surcharges

Surcharges are not considered taxes, and even tax-exempt organizations must pay them. Adding surcharges to your telephone bill is one way state legislators raise revenue without technically raising taxes. Surcharges that effect long distance service are covered in more detail in the chapter on long distance bill auditing.

Case Study # 12

GTE – Texas based customer receives $360,000.00 refund for incorrectly calculated emergency service surcharges.

A large data processing firm located in Texas was obligated to pay a per line monthly charge of $0.75 imposed by the local municipality, as an

Emergency Service Charge. We contacted the municipality and there was a stipulation that this fee would be billed only to the first 100 lines. This firm however, had 20,275 Centrex lines incurring this charge for an extended period of time. After an audit and confirmation of the billing error, this client received a refund of $360,000.

Case Study # 13

SNET – Connecticut based customer receives $120,000.00 refund for taxes billed in error.

An Indian Reservation in the northeast was entitled to an exemption from federal and state taxes. They were erroneously billed taxes for over two years. After an audit and confirmation of their tax-exempt status, they received a refund of $120,000.

Case Study # 14

GTE – Texas based customer receives $160,000.00 refund for federal tax billed on Centrex service in error.

Some components of Centrex service are federal tax exempt. GTE's billing system was applying federal tax to all parts of Centrex service.

CABS errors

Refunds for CABS billing errors are limited to two years. The following are some examples of CABS billing errors:

Case Study # 15

Southwestern Bell – Texas based customer receives $24,000.00 refund for S25 surcharge billed in error.

The CABS system is designed to automatically bill you an S25 surcharge unless you provide them with an exemption certificate. The S25 applies only if a point to point circuit somehow accesses the switched network. Since this is rarely the case, most CABS customers provide a blanket exemption to the LEC. An audit found that three T1s were billed the S25 in error.

Case Study # 16

GTE – Texas based customer receives $150,000.00 refund for tariff error.

GTE's tariff called for a discount to be applied to each additional T1 in service connecting the same locations. This customer had over 48 T1s between the same two locations and all T1s were charged the full rate.

Figure 4.15

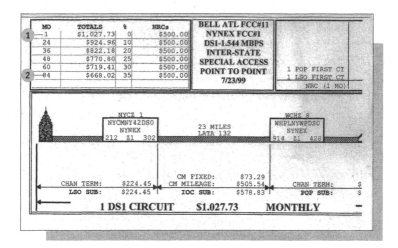

Case Study # 17

Bell Atlantic – NY based customer receives $35,000.00 refund for term plan error.

An audit of a CABS bill found 4 T1s connecting locations in New York City to White Plains NY billed at a monthly rate of $1,027.73. A contract was found whereby the customer agreed to keep these circuits for seven years (84 months) in return for a discount of 35% off the standard tariff rate ($666.02). Figure 4.15 shows the difference in the pricing where 1 is the standard rate and 2 denotes the 84 month discount rate. The difference in the standard rate and the discount rate had to be credited on all four T1s back two years. Figure 4.15 is a hardcopy from the LATTIS software program. We use LATTIS to price all our CRIS and CABS circuits, verify interoffice mileage and term discounts.

Initiating an Overbilling Claim

Once you identify an overbilling error such as those listed above, how do you go about obtaining a refund?

- Gather all documentation you have. Check with your equipment vendor to obtain information they may have on moves and disconnects.

- If you find that a circuit or telephone line in question was disconnected, try to determine the actual disconnect date. Be careful here, the activity date listed on the CSR may not be the date the circuit should have been disconnected. The activity date field will change anytime a rate change is approved that effects its associated USOC. Many times, LEC employees will tell you that a certain circuit could not have been disconnected on the date you claim it was. As proof they will point to the activity date and say someone must have requested a change to this circuit. You must challenge this if you are convinced their conclusion is incorrect.

Try to find associated USOCs that are not billed a monthly rate to establish your case. The following example illustrates how this can be accomplished.

QUANT	ITEM	DESCRITPTION	ACT DATE	RATE
1	PFSAS CKL	/CKT 96,BANA,,99916,,NY 1-444 BROADWAY, MANHATTAN NY/DES FLR 3/LOS 212 966/CLS 96,BANA,,99916,,NY	09-22-87	
1	CON2X	/EAR (LOOP CHARGE 2-WIRE)	02-1-90	21.53
1	NRASQ	/DES 42A BLK HOLMES INSTR SOUTH STAIRWELL		
1	PMWV2	/EAR (BASIC 2-WIRE VOICE CIRCUIT)	02-01-90	7.07

In this case, the customer told us that this circuit was disconnected on 09-22-87. When we initially contacted the LEC, their representative responded that this was impossible. The reason given was that the ACT DATE (activity date) listed on the CSR was 02-01-90. What the representative did not know is that Bell Atlantic -NY had a massive rate change on 02-01-90 that affected most special service circuits and therefore caused the ACT DATE to change.

We established that the USOC PFSAS (which lists the Circuit ID) was a better indicator of the last time physical activity occurred on this circuit. As there is not a monthly rate for this USOC, it is not effected by rate changes. The Moral? Do not assume information given by LEC representatives is correct. Some representatives are better trained and more knowledgeable than others.

The following example shows how a typical overbilling case develops: Assume that after diagramming the locations of all your circuits you find that you are still being billed for off-premise outside extensions (OSNA circuits) to a branch location closed down five years ago. Your problem is that you can't find any records that prove you called your LEC to disconnect them. What can you do?

Step 1 - Gather together all the information you have on the move. Look for anything that backs up your position and establishes the date you moved out of your branch office. For example, you may be able to get a copy of your old lease from your real estate agent or landlord. You may be able to locate moving bills from a trucking company that identifies the date you vacated the premise in question. Look for any bill, record or lease that logically proves you were unable to use the circuit because you did not have access to the building where the circuit in question terminated.

Step 2 - Call the LEC's billing office and identify the circuits in question. Tell them the date you moved out of the building in question. Be prepared to overcome objections such as "how were we suppose to know you moved?" Be persistent! Request a supervisor if you feel that you are getting the third degree. Eventually your claim will probably be sent to a specialist in overbilling claims. These special representatives investigate and correct overbilling problems. Put your claim in writing and send it to the specialist that is assigned to your case.

Step 3 – Have the LEC send a technician to verify that the circuit is no longer in service (known by the LEC as a physical inspection). They should test the circuit to see if it is still able to function by putting a tone generator on the originating end and then check the terminating end to see if they can pick up the tone. The logic behind this test is to determine if the circuit is usable. If the circuit is found to be non-operational you should request a refund where state tariffs allow or out of service credit. You will need to negotiate with the LEC to determine the amount of backcredit.

The Secret to Negotiating Refunds

Some state tariffs are more favorable to consumers than others. Many states have statutes of limitation laws that limit the number of years companies are required to give credit, even when bills are proven to be wrong. The attitude of some LECs is "Consumer Beware". They know their systems are riddled with errors, but unless you find the errors quickly its tough luck.

As competition slowly seeps into the local market some LECs are becoming more customer focused. Those companies located in more competitive markets like New York City, Chicago and Los Angeles will issue credits and refunds when wrong as customers have more alternatives. Our advice is to be persistent. You may need to speak to an executive that understands the bottom line and has the authority to help you.

Remember that the LEC representative that initially denies credit is near the bottom of the LEC hierarchy. This representative operates under strict guidelines. They can only offer credit in limited situations. They are also limited as to the amount of a refund they can offer. The hierarchy of a typical LEC is as follows:

Level	Job Title
Non-Management	Representative
First	Supervisor
Second	Manager
Third	District Manager
Fourth	Division Manager
Fifth	Asst. Vice President
Sixth (officer)	Vice President
Seventh (officer)	Chief Operating Officer
Eighth (officer)	President

If you are not satisfied with the credit offered by the representative, ask to speak with a supervisor. Each higher level at the LEC has more leeway than the previous level to grant credit. You have a chance of being more successful if you present your case at a higher level.

Contacting the State Public Utility Commissions

You should only refer problems and disputes to state regulatory agencies as a last resort. Before you lodge a complaint, you should at least take your case to the sixth level manager (officer level) at your LEC. Make it clear to the LEC that you are planning to lodge a formal complaint. It will help your case. Many regulatory bodies are moving toward procedures whereby rate hike approvals are dependent upon the quality of service provided to LEC customers. Because of this trend many LEC executives are rated on how many public complaints are registered against their department. These ratings can affect their year-end bonuses.

How to Determine if you Should Use A Professional Bill Auditor

How much time do you have to spend on the auditing process? Performing an audit can be a time consuming, frustrating process that does not always lead to a refund. How complex is your bill? Is it cost effective for you or your people to learn USOCs? The day to day problems in running a telephone department can be overwhelming. A professional auditor should review your bills on a contingency fee basis. If they do not produce refunds then they do not get paid.

How to Select an Auditor

Selecting an auditor should not be any different than selecting a lawyer or other professional. Competency and trustworthiness are essential. You are going to empower the auditor you select to examine all your records, interface with your co-workers and vendors, and negotiate with LECs on your behalf. Check the auditor's background and references.

When interviewing auditors, question if they require you to make copies of all your bills. This can be quite time consuming. The auditor should be required to make their own copies of what they need. All copies should be made on your premises.

There is also a subjective side to selecting an auditing firm. Do you feel comfortable dealing with them? Remember you are authorizing them to act on your behalf. You do not want someone who is loud and obnoxious representing your company. Choose wisely!

Telecommunications Expense Management

5

Optimizing
Your
Network

Optimization is the process of analyzing your telecommunications network to find products and services that will reduce your monthly bill and increase network efficiency. Optimization should be risk free and easy to implement. Optimizing your network is largely implementing ideas that "make sense".

Saving money can be as simple as calling the Telephone Company to convert a particular service to a Term Plan or having your IXC transport your local calls. The possibilities are endless but the savings are real.

Sample Optimizations

The following lists some common Optimizations:

1. Implement LEC local toll call discount plans. LECs offer a multitude of local discount plans. You will need to evaluate these plans in order to find the most cost efficient plan for your volume and calling needs.

    As more IXCs now offer intraLATA toll calling, the LECs are responding with reduced local toll rates. For example, Pacific Bell offers discount-calling plans known as Advantage 5, 10 and 25. These plans can save you up to 50% on your local calls. Advantage 25 provides large users with a per minute local call rate of 3.6 cents a minute. Ameritech offers a discount-calling plan called Netspan and Bell Atlantic has a multitude of plans as well.

2. Compare your IXC local call rate with the LEC local discount plan rates. Switching your local toll calls within a LATA to your IXC is easy. Each telephone line now has an LPIC (local PIC) associated with it. Both the PIC and the LPIC can be switched to your IXC. You are free to choose your IXC for local calling in most areas.

3. Install a T1 to your carrier's POP. Your IXC will significantly reduce your long distance rates.

4. Sign a long-term commitment to keep a particular telephone company service and they will usually provide you with a lower monthly rate. These plans are most commonly applied to special service and CABS circuits. They make sense if you own your building(s) or have signed long term leases and plan to stay at the same location(s) for an extended period of time. A five-year term plan will usually provide a monthly discount of about 35% off the standard tariff rate.

5. If your company has many BTNs, have the LEC combine them onto one summary bill with one payment date. Your company will save the cost of processing multiple bills and sending out multiple payments. This will also help your company avoid late payment charges, as you will only have to pay by the due date on that one account rather than the due date on each of the individual accounts.

6. Perform a physical inspection of all of your facilities. Identify and disconnect unused facilities.

7. PBX Optimization Opportunities. Your PBX can help you identify underutilized and unused trunks. This information can be obtained from your PBX vendor or via modem access. The following steps should be taken:

a) Review PBX traffic reports to identify PBX trunk utilization.

b) Compare PBX reports with the facilities billed on the CSR. Generate Optimization reports to remove unused facilities and PBX equipment (trunk cards, T1 cards, extension cards etc).

8. Make sure all long distance calls are being billed by your primary IXC at your custom contract rate.

Case Study #1

This company has its corporate headquarters in Chicago with a distribution division that has 150 small branch offices throughout the country. Corporate negotiated a national contract with AT&T for all of the company's various divisions and subsidiaries. As a cost savings measure an auditing firm was hired to perform an Optimization on the 150 distribution branches.

A copy of every CSR, local and long distance bill was audited. Each bill was analyzed for the following:

1. Long distance calls billed by any IXC on the local bill. These calls would be billed at an IXC's highest rate. If these calls were found, the PIC was changed to ATI. (Note AT&T can be PICed as either AT&T or ATI. The PIC AT&T will generate AT&T's standard tariff rates while ATI will generate custom tariff rates).

2. That all long distance bills were part of the AT&T contract. If not, they were switched to AT&T.

3. The LECs best local calling rates for 800 and outbound service were compared to AT&T's local rates. The calls were either switched to the LEC's best local plan or to AT&T's custom contract rate.

The following spreadsheet was put together for each branch office.

Services	Current Cost	Proposed Projected Cost/Month	Proposed Annualized Cost Savings
Monthly Cost (Trunks/Circuits)	$2,472.16		
"800" Interstate Minutes	1,990.18		
"800" Interstate Cost	$187.08		
"800" Interstate Cost/inute	$0.0940		
"800" Intralata Minutes	5,635.06		
"800" Intralata Cost	$529.70		
"800" Intralata Cost/Min	$0.0940		
Outbound Interstate Minutes	12,216.24		
Outbound Interstate Cost*	$1,165.17		
Outbound Interstate Cost/Min*	0.0954		
Outbound Intralata Minutes	5,635.06		
Outbound Intralata Cost	$556.10		
Outbound Intralata Cost/Min*	0.0987		
Calling Card Minutes	5,244.00		
Calling Card Cost*	$492.94		
Calling Card Cost/Min*	0.0940		
Conference Service Minutes	500.00		
Conference Service Cost*	$400.00		
Conference Service Cost/Min*	$1.35		
Special Circuit Costs*	$18.04		
International Call Traffic Minutes	200.00		
International Call Traffic Cost	$180.00		
International Call Cost/Min*	0.90		
Projected Monthly Savings	$1,605.00		
Projected Annual Savings	$19,260.00		

"Recommendations/Comments: IntraLATA calls billed on AT&T VTNS. Paying for fixed cost for a SBC discount intraLATA savings plan with no usage."

The result of this Optimization was an average yearly saving of $20,000.00 for each branch. The Optimization resulted in yearly savings of $3,000,000.00 for that division.

Case Study #2

A large bank with over 200 branches was a heavy user of Centrex from the LEC. Each branch obtained its Centrex service from the LEC individually at a cost of $26.32 per Centrex line. Each branch was billed an average of $300.00 for local calling at an average rate of 10 cents per minute.

A statewide Centrex contract and local calling discount was negotiated with the LEC. The Centrex per line rate was reduced to $9.75 per line and the local calling rate was reduced by 40% to 6 cents per minute. Optimization savings of $900,000.00 were calculated for these simple changes calculated as follows:

2400 Centrex lines @ $26.32 per line	$ 63,168.00
2400 Centrex @ $9.75	$ 23,400.00
Monthly savings	$39,768.00
Total cost for local calls	$ 60,000.00
40% discount	$ 24,000.00
Monthly savings	$ 36,000.00
Total monthly savings	$ 75,000.00
Total yearly savings	$900,000.00

Case Study # 3

This New York based company had a total of 600 analog DID trunks at various locations. It had installed digital PBXs but continued to utilize analog DID trunks. As part of an Optimization study, the cost effectiveness of digital switched T1s (brand name: Flexpath) was evaluated.

The 600 DID trunks could be provided as digital channels over switched T1s. 25 switched T1s (600 divided by 24) would be required.

The monthly savings on each T1 was calculated as follows:

Step 1 –The full cost of an analog DID trunk was calculated as $83.38.

TB2 (DID Trunk Charge)	$56.17
D1F2X (loop charge)	$19.08
9ZR (FCC line charge)	$ 8.13
TOTAL	$83.38

Step 2 - Determine the monthly cost of 24 DIDs. The monthly cost for 24 DID trunks is $2001.12 (24 times $83.38).

Step 3 - Determine the monthly cost of a T1. You must also determine the installation cost of the T1 so you can calculate the break-even point. Installation and monthly charges for a single Bell Atlantic - NY switched T1 are calculated as follows:

	Monthly	**Installation**
Service Order		$ 56.00
Premise Visit		$ 19.00
Group of 24 ports	$ 533.06	
Digital Transport Facility	$ 435.87	
24 FCC line charges @ $8.1	$ 131.52	$ 1,550.00
TOTAL	$1,100.45	$1,625.00

Step 4 - Determine other conversion and installation costs. To utilize the T1, you will need a Channel Service Unit (CSU). A CSU regenerates digital signals and monitors your T1 for problems. In addition, your PBX will probably need a T1 card to connect the T1 to the PBX. In this case the cost for the CSU and T1 card was $5,000.00.

Step 5 - Add the PBX installation costs and the LEC's installation costs to determine the total cost to convert to a T1. $1,625.00 plus $5,000.00 equals $6,625.00.

Step 6 - Calculate the break-even point by dividing the installation costs ($6,625.00) by the monthly savings ($900.67). Your break-even point is 7.4 months. After the first 7.4 months your company will save $10,808.04 a year per T1.

Step 7 – After the breakeven point of 7.4 months, the cost savings for 25 switched T1s is $270,201.00 a year.

6

Long
Distance
Bill
Auditing

Understanding Your Long Distance Chargers

There are two ways to access your long distance carrier's network. One way is to gain access through your LEC's CO (switched access). With this method, your call travels over your existing telephone lines to the local LEC CO. There your call is then "switched" to the long distance network.

The second method of access called Dedicated Access through a dedicated circuit, like a T-1 link, that connects you directly to your long distance carrier's Point-of-Presence (POP) without being switched by the local CO. This method is sometimes called "bypass" because your call does not "pass through" the LEC's CO.

There may soon be a third way to access long distance networks. Access is through voice over IP. This basically uses a data network and the TCP/IP computer networking protocol (the same one used by the Internet) to send voice calls from one place to another.

Packet switching for voice has its problems. Voice is too sensitive to latency, or delay, especially the variable delay associated with packet-switching networks. Recently developed software and DSP hardware that uses sophisticated compression techniques have improved the ability to conduct "reasonable" quality packet voice conversations over the Internet.

LECs used to charge long distance carriers a usage sensitive access charge for each and every call switched through their CO. Approximately 25% of the cost of a long distance call stemmed from this usage sensitive cost component.

This changed with the Telecommunications Act of 1996. The usage sensitive portion was eliminated, and as of January 1, 1998, local phone companies now charge long distance carriers a flat Presubscribed Interexchange Carrier Charge (PICC). This is a monthly per line cost charged by the local LEC to each long distance carrier for every customer phone line presubscribed to that carrier.

Long distance carriers typically pass this charge onto their customers. In many cases, they pass it off as a surcharge, but it really isn't since they are not obligated to impose it.

The PICC per month, per line charges are as follows:

Business:

Single Line --	$0.53
Multi – Line --	$2.75
Multi – Line (CA only) --	$1.84
Centrex (Single Line) --	$2.75
Centrex (2 – 8 lines) --	$0.72
Centrex (9 or more Lines) --	$0.31

Residential:

Single Line (Primary) --	$0.53
Additional Lines (Non-Primary) -	$1.50

The Audit - Getting Started

In order to get started you will need to obtain a copy of your IXC contract, all contract amendments and a copy of at least three months of bills (both paper and CD-ROM disks). You will also need to obtain a copy of your IXC Management Reports (summary of calls by location), Call Detail Data Reports (listing of calls) and access to your IXC's tariffs (via an online commercial tariff service such as Telview). It is also helpful to obtain a copy of your IXC Virtual Network Routing Tables. These tables list the NPA/NXXs of all the sites that should be considered "on-net". If you do not currently receive these reports, request them from your AE.

Using IXC Tariffs to Audit Your Bills

IXCs must file tariffs with both the FCC and with every state's PUC in which they conduct business. The FCC regulates the following services:

• Interstate Calling & Calling Plans
• International Calling Rates
• Calling Cards
• 800 Calling
• Virtual Networks (VPN, VTNS, VNET)
• Access Circuits (Dedicated circuits to POP)
• Leased Line Circuits (interstate)
• Frame Relay (interstate)
• A repository of Individual Case Basis (ICB) contract tariffs

Intrastate services are filed with each state's PUC. Rates and regulations vary by state. Services tariffed with the state include intrastate interLATA calling and intrastate intraLATA calling.

Figure 6.1 lists the different carrier tariffs available for viewing online with Telview's online commercial service. Though many tariffs are online free of charge, you may still require a commercial online tariff service. Figure 6.2 lists some of the AT&T tariffs Telview has available online while Figures 6.3 and 6.4 list the various MCI and Sprints tariffs Telview has available online.

Figure 6.1 Resource Required Carrier List

To view tariffs, first choose a carrier from the following list.

AT&T	MCI	Sprint
Access Transmission	GTE Communications Corp	RCI Corporation
ACSI	GTE Tele Oper. Co.	Rochester Telephone
Affinity	Gulf Telephone	Roseville Telephone
ALASCOM	GVNW	RTMC & NY RSA No.3
Alaska Tel Util	Hargray	Seneca-Gorham Telco
Aliant	Horry	SNET America, Inc
Allegiance Telecom of Texas, Inc	ILL Consolidated Tel	Southern New England
ALLTEL	ILL Small Co ECA	SouthernNet
ALLTEL Communication	Illinois Bell	SouthWestern Bell
Ameritech	Illinois Metro	Sprint
AMNEX	Indiana Bell	Sprint Comm. L.P. Long Distance
Appalachian Cellular	IT&E Overseas, Inc	Sprint Local Tele Co
AT&T	ITI D/B/A ONCOR	St. Joe
Ausable Valley Telco	LCI International	SW Bell Comm Svc
Bell Atlantic	LDDS	Sylvan Lake Telco
Bell South	Lincoln Telephone	Taconic Telephone Co
Blue Ridge Cellular	Loral Skynet	TCG
BOCS	Lufkin-Conroe Tel	TCI
Brooks Fiber Communications	Mankato	Telco
Cable & Wireless USA, Inc	MCI	Teleconnect Phone Co
Cable & Wireless USA, Inc	MCI Metro Trans Service	Teledial America Inc
Carolina Tel & Telgr	MCLEODUSA	TelTrust, Inc
Centel	Metromedia	Time Warner
Century	MFS Intelenet Inc	Timer Warner Connect

AT&T	MCI	Sprint
CenturyTel Operating Companies	MFS International Inc	TLDPR
Cinn Bell	MFS Telecom Inc	TSTCI
Cinn Bell Long Distance	MI Exchange Carr Asc	Tueca
Citizens	Michigan Bell	U.S. Long Distance
CleartTel Comm. Inc	N. Carolina Telcos	United
CNSI	NECA	US Networks
Coast To Coast Telecommunication	Nevada Bell	US West
Com Systems, Inc	New England Tel	VA Telecom Indst. Assc
Conquest Oper Svcs	New York Telephone	Vitel Co
Eagle Telecom - Co	NextLink	VYVX
EMI	NY Settlement Pool	Warwick Telephone Co
Excel	NY State Tel Assoc	Washington Tel Util
Frontier Comm	NYNEX	Western Union
Frontier Company	OCI D/B/A ONCOR	WI State Tel Assoc
Frontier-Rochester Telecom	Ohio Bell	WILTEL
GCI Communication	Oreca	Wisconsin Bell
Gem State Utilities	Oregon Tel Utilities	WorldCom
GSTC	Pacific Bell	WorldCom Network Services, Inc
GTC, Inc	Pita	WorldCom Technologies, Inc
GTE - (Contel)	Puerto Rico Telephone Co.	
GTE Card Service	Qwest Communications Corp.	

Telview– List of On-Line Tariffs
http//:www.Telview.com

Figure 6.2   Resources Required AT&T Tariff List

Federal Tariffs

View		Download
FCC #1	IS Long Distance MTS	(10170KB)
FCC #2	IS Wats Svc	(1893KB)
FCC #4	Switched Digital	(549KB)
FCC #5	Special Construction	(155KB)
FCC #7	Skynet Satellite Rate Tab	(1446KB)
FCC #7	Skynet Satellite	(768KB)
FCC #9	Private Line	(3665KB)
FCC #9	Private Line Rate Tables	(7768KB)
FCC #10	Mileage Info & Admin	(2633KB)
FCC #11	PL - Local Channel	(5574KB)
FCC #11	PL - Local Channel RT	(68570KB)
FCC #12	Custom Des. Integ. Svc	(12820KB)
FCC #13	Overseas Svcs	(830KB)
FCC #14	Wats (PR & Virgin Isl)	(105KB)
FCC #15	Competitive Pricing	(98KB)
FCC #16	Competitive Govt Svc	(3519KB)
FCC #16	Competitive Svc APDX	(51396KB)
FCC #16	Competitive Svc Rt	(1345KB)
FCC #17	Telex Svc	(103KB)
FCC #18	Teletype Exchange Svc	(78KB)
FCC #19	Worldwide Telex Svc	(53KB)
FCC #20	International Telegraph	(158KB)
FCC #21	Dom Telex-O/S Interntl	(42KB)
FCC #22	Dom TWX - O/S Interntl	(32KB)
FCC #23	Dom Telex-Dom Com Car	(39KB)
FCC #24	Dom TWX - Dom Com Car	(37KB)

Federal Tariffs

View	Download
FCC #25 International Service	(71KB)
FCC #26 Packet Switched Svc	(98KB)
FCC #27 Consumer Telecom Svc	(5504KB)
FCC #28 Access Service	(212KB)
FCC Contract Tariff Volume 46	(693KB)

State Tariffs

View	Download
AL - A General Services	(696KB)
AL - B Custom Network Svc	(952KB)
AL - C Channel Service	(4KB)
AL - D Digital Service	(5KB)
AL - F Switched Digital Svc cancelled	(193KB)
AL - G Private Line	(767KB)
AR - 9 Private Line	(431KB)
AR - 11 Private Ln Local Chan	(576KB)
AR - Custom Network Svc	(546KB)
AR - Digital Data Svc	(144KB)
AR - Local Exchange Service	(119KB)
AR - Message Telecom Svc	(1152KB)
AR - Switched Digital Svc cancelled	(3KB)
AR - Wats	(72KB)
AZ - Admin Tariff Practice	(138KB)
AZ - 11 Private Ln Local Chan	(661KB)
AZ - 9 Private Line Services	(479KB)
AZ - Telecom Svc Tariff	(1233KB)

State Tariffs

View	Download
CA - CPUC #A-T	(2818KB)
CA - CPUC #B Private Line	(214KB)
CA - CPUC #C 1-T	(195KB)
CA - CPUC #C 4-T Switched Dig Svc	(660KB)
CA - CPUC #C 5-T Special Const	(83KB)
CA - CPUC #C9-T Private Line	(1147KB)
CA - CPUC #C10-T Mileage Info	(5KB)
CA - CPUC #11-T Priv Ln Local Chan	(1070KB)
CA - CPUC #E-T Competitive Local Car	(461KB)
CO - Private Line	(475KB)
CO - Private Ln Local Chan	(630KB)
CO - Local Exchange Service	(147KB)
CO - PUC #2 Access & Network Interconnection	(458KB)
CO - Telecom Svc Tariff	(729KB)
CT - DPUC #1 Custom Netwk Svc	(1334KB)
CT - DPUC #2 Fed Telecom System	(12KB)
CT - DPUC #3 Switched Digital Svc cancelled	(106KB)
CT - DPUC #4 Ld Service	(801KB)
CT - DPUC #5 Local Exchange	(280KB)
DE - PSC #2 Intrastate Tariff	(585KB)

Figure 6.3 Resources Required MCI Tariff List

Federal Tariffs

View		Download
FCC #1	Customized Bus APDX	(37684KB)
FCC #1	Customized Business	(10326KB)
FCC #1	International Service	(416KB)
FCC #6	Information Tariff	(19KB)
FCC #7	Govt Telcom Svc	(2259KB)
FCC #8	Dedicated Acc Line Svc	(14028KB)

State Tariffs

View		Download
AL -	APSC #1 Intercity Telcom Svc	(846KB)
AR -	APSC #1 Intercity Telcom Svc	(1055KB)
AZ -	ACC #1 Intercity Telcom Svc	(1193KB)
CA -	CPUC #1T-7T Telcom Utility Sv	(1235KB)
CO -	Price List #1 Intrastate Telco	(785KB)
CT -	DPUC #1 Intrastate Telcom Svc	(488KB)
DE -	PSC #1 Intrastate Telcom Svc	(638KB)
FL -	FPSC #1 Price List	(26KB)
FL -	FPSC #2 Intercity Telcom Svc	(1025KB)
GA -	GPSC #1 Intercity Telcom Svc	(1043KB)
HI -	PUC #1 Intrastate Telcom Svc	(451KB)
IA -	#3 Intrastate Telecom Svc	(1122KB)
ID -	Price List #2 Intrastate Telc	(1057KB)
IL.	C.C. #3 Interexchange Telcom	(1064KB)
IN -	I.U.R.C.#T - 2 Intercity Telcom	(857KB)
KS -	CC #1 Intercity Telcom Svc	(1462KB)

State Tariffs

View	Download
KY - KPSC #1 Intercity Telcom Svc	(877KB)
LA - LPSC #1 Intercity Telcom Svc	(967KB)
MA - DPU #3 Intercity Telcom Svd	(803KB)
MD - PSC #3 Intercity Telcom Svc	(876KB)
ME - PUC #1 Intrastate Telcom Svc	(578KB)
MI - M.P.S.C. #1R Interexchange Te	(429KB)
MN - Price List #1 Intercity Telco	(906KB)
MO - MPSC #1 Intercity Telcom Svc	(1067KB)
MS - MPSC #1 Intercity Telcom Svc	(1626KB)
MT - Tariff #2 Intrastate Telcom S	(434KB)
NC - NCUC #1 Intercity Telcom Svc	(935KB)
ND - Price List #2 Intrastate Tel	(816KB)
NE - Price List #3 Intrastate Telc	(836KB)
NH - PUC #1 Intrastate Telcom	(446KB)
NJ - BPU #3 Intercity Telcom Svc	(767KB)
NM - CIT #1 Intercity Telcom Svc	(956KB)
NM - No.2 Virtual Ntwk Telcom	(284KB)
NV - PSCN #1 Intercity Telcom Svc	(1062KB)
OH - P.U.C.O. #1 Intrastate IX To	(1165KB)
OK - OCC #1 Intercity Telcom Svc	(2336KB)
OR - Price List #2 Intrastate Tel	(960KB)
PA - PUC #2 Intercity Telcom Svc	(850KB)

Figure 6.4 Resources Required SPRINT Tariff List

Federal Tariffs

View		Download
FCC #1	Cellular Service	(82KB)
FCC #1	International Service	(50KB)
FCC #1	Interstate Services	(16961KB)
FCC #2	IS Wats Svc	(4115KB)
FCC #3	Analog/PL Svc	(174KB)
FCC #4	Govt Telcom Svc	(32111KB)
FCC #5	Virtual Private Ntwk	(955KB)
FCC #6	Video Telcom Svc	(309KB)
FCC #7	Digital Private Line	(1043KB)
FCC #8	Access & Misc Svcs	(8276KB)
FCC #10	Offshore Comm Carrier	(816KB)
FCC #11	Business Comm	(16802KB)
FCC #12	Custm Netwrk Svc Arrg	(16383KB)

State Tariffs

View	Download
AL PSC #3	(516KB)
AR PSC #2	(555KB)
AZ CC #2	(998KB)
CA PUC #1	(982KB)
CO Price List #2	(478KB)
CT - DPUC #3	(501KB)
DE PSC #2	(501KB)
FL PSC #2	(913KB)
GA PSC #3	(621KB)

State Tariffs

View	Download
HI PSC #1	(436KB)
IA SCC #3	(445KB)
ID PUC #3	(573KB)
IL CC #5	(688KB)
IN - IURC #2	(865KB)
KS SCC #3	(661KB)
KY PSC #4	(747KB)
LA PSC #3	(572KB)
MA DPU #3	(641KB)
MD PSC #3	(551KB)
ME - Rate Schedule	(896KB)
MI PSC #3R	(387KB)
MN PUC #2	(570KB)
MO - Competitive Services	(91KB)
MO PSC #2	(1018KB)
MS PSC #4	(1010KB)
MT - Price List No. 1	(518KB)
NC #2	(744KB)
ND PSC #2	(439KB)
NE PSC #3	(514KB)
NH PUC #4	(532KB)
NJ BPU #2	(549KB)
NM SSC #1	(913KB)
NV #2A (Central) General Customer Services	(2742KB)
NV #3A (Central) Intrastate Access Services	(2094KB)
NV PSC #1B	(801KB)

Resources Required SPRINT Tariff List (Continued)

State Tariffs

View	Download
NY PSC #2 - Telephone	(1567KB)
OH PUCO #2	(711KB)
OK CC #2	(596KB)
PA PUC #2	(817KB)
RI PUC #2	(520KB)
SC PSC #3	(974KB)

Sprint Internet Web Site Consultant Liaison
http//:www.sprintbiz.com/consult
(Access Sprint Rates Including Intrastate Rates)

An audit of your IXC bills should be divided into two major work efforts or categories:

I. Voice Services: This category includes interstate (across LATAs in different states, intraLATA (local calling within a LATA), intrastate (between LATAs within a state) and international calling.

II. Leased Line (Access) Circuits: This category includes T-1 circuits used to carry voice traffic to the IXC POP, leased line circuits (point to point-outside the LATA), analog (4.8kb, 9.6kb) circuits, digital (DS0, DDS, T-1, T-3) circuits, and frame relay.

I. Auditing Voice Services

When auditing your voice services you need to be aware that different rates apply to calls that are from or to "on-net locations". An on-net location is a company location that has a T-1 connection to the IXC's POP (dedicated access). An off-net location does not have T-1 access to the

POP (switched access).

Calls originating from any of your various locations will fall into one of the following rate categorizes:

Originating Loc.	to	Terminating Loc.	Rate Category
On-Net Location		On-Net Location	On to On
On-Net Location		Off-Net Location	On to Off
Off-Net Location		On-Net Location	Off to On
Off-Net Location		Off-Net Location	Off to Off

Calls are billed by the IXC for each rate category according to the rates listed in your contract. If you install T-1s at any of your locations you must now verify that calls made from and to that location are billed at the proper rate. Most large companies will have a combination of both switched (off-net) and dedicated (T-1 access) at different locations. When your T-1s are filled to capacity, voice calls will overflow onto your switched network resulting in a higher rate per minute charge. If this traffic is significant you should increase the number of T1s at that location. Figure 6.5 depicts a typical voice network configuration.

Figure 6.5 Voice Service Billing Review
IXC's LD Network

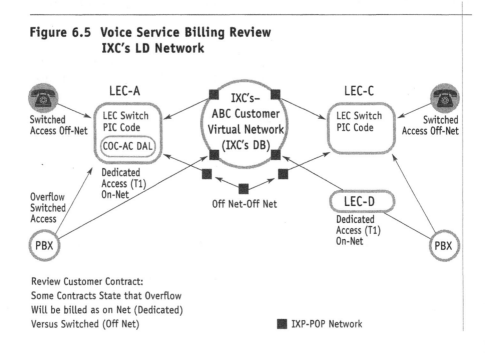

Your IXC may not properly update its Virtual Routing Tables to flag the location as an on-net location. Your Virtual Routing Tables list the NPA/NXXs that are on-net. You have to check your bills and verify that originating and terminating calls are being rated under the proper rate category. Figure 6.6 details MCI's internal VNET (Virtual Network) architecture.

Figure 6.6 MCI's VNET Architecture

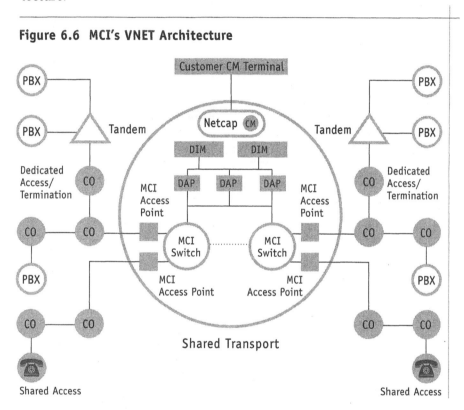

Figure 6.7 is a copy of a Sprint Customer Contract. The contract lists the per minute prices for each rate category. It also lists a postalized rate for Canada of 29 cents from an on-net location and 32 cents made from an off-net location. A postalized rate is a flat rate per minute. As part of our audit we will want to verify that the proper domestic rate categories are applied on calls to Canada from both on-net and off-net locations.

Figure 6.7

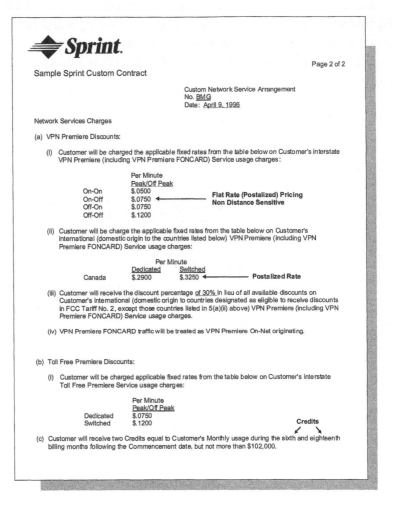

The Sprint contract also states the customer will receive two credits equal to customer's monthly usage during the sixth and eighteenth month not to exceed $102,000.00. One of the most common errors uncovered in audits is that IXCs often fail to provide these type of credits.

Figure 6.8

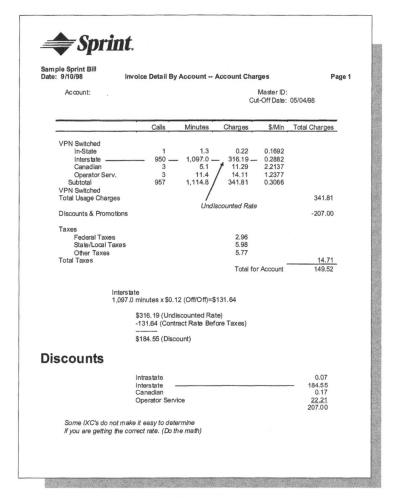

Most IXC bills are not user friendly and that makes verifying your rates difficult. This contract has an off-to-off postalized rate (switched) of 12 cents. Figure 6.8 is page 1 of the bill sent by Sprint to this customer. In order to verify the 12-cent rate we need to check the VPN switched rate. 1,097 minutes were used. Since we should be billed at a rate of 12 cents, the total cost should be 1,097 minutes times .12 cents or $131.64.

However, the charge listed for switched calls is $316.19. We must check the bottom of page 1 where Sprint lists the discounts applied to the bill for switched calls. Instate calls have a discount listed of $184.55. If we subtract $184.55 from the $316.19 billed we come up with a total of $131.64. $131.64 divided by 1,097 minutes equals the contract rate of 12 cents.

The more complicated the contract the more an IXC has to manipulate its billing system to meet the discounts agreed to in the contract. An IXC will most often have your calls rated at the prevailing tariff rate. Then they will have line item discounts at the end of the bill that will discount calls so they meet the contract rate. IXCs can also hardcode the contract rate at the call detail level. In the example above, Sprint could have hardcoded the 12 cents rate into its bill system at the call detail level but chose not to. In complicated contracts an IXC will often have a combination of line item and hardcoded rates at the call detail level. The more complicated the contract, the greater the opportunity for errors.

When verifying your rates you also need to take into account various off-peak discounts. These discounts are listed in an IXC's tariffs. Figure 6.9 lists the discounts in place for AT&T's SDN service. You also need to take into account how your IXC performs rounding. Most IXC will round a call that is 4.4 cents down to 4 cents and round up a call for 4.5 cents to 5 cents.

Your contract may protect you from the fees introduced by the IXCs as a result of the Telecommunications Act of 1996. Many contracts have clauses that prohibit the addition of surcharges during the life of the contract. MCI's tariff actually prohibits the billing of the PICC charge on all custom contracts (SCAs) signed prior to January 1, 1998.

The following is a summary of the surcharges and fees introduced as a result of the Telecommunications Act of 1996.

- Payphone Surcharge: The IXCs add a 30 cents surcharge to calls that originate from payphones. IXC Call Detail Records (CDR) will indicate the call came from a payphone. You will see this surcharge most commonly on calls made via credit card.

Figure 6.9 Rate Periods

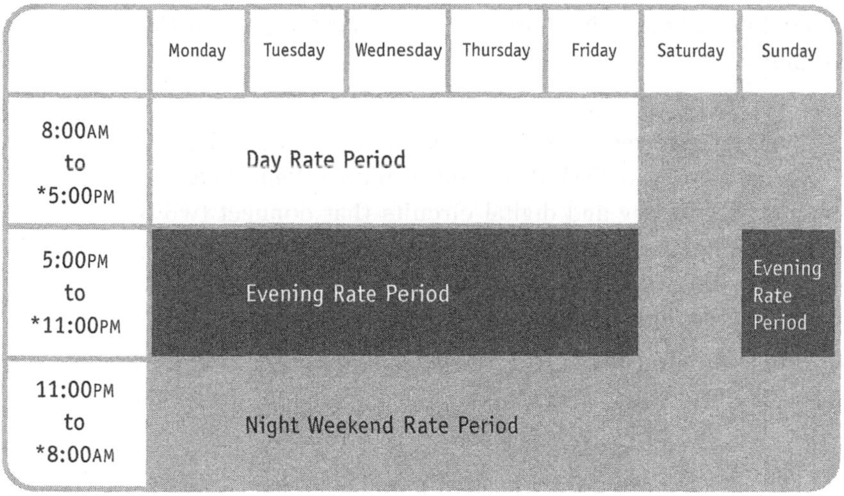

	Monday	Tuesday	Wednesday	Thursday	Friday	Saturday	Sunday
8:00AM to *5:00PM	Day Rate Period						
5:00PM to *11:00PM	Evening Rate Period						Evening Rate Period
11:00PM to *8:00AM	Night Weekend Rate Period						

*To but not including

- Presubscribed Interexchange Carrier Charge (PICC): PICC surcharge applies to every access line (including DID trunks). If the access line does not have a PIC selected, then the LEC will bill the PICC.

  - Single Residence Line: $0.53 (Sprint bills $0.85)

  - Multi-Line: $2.75 (AT&T bills $2.50)

  - Centrex Line (over 9): $0.31

- Each IXC has its own name for the PICC: MCI refers to the PICC as the National Access Fee (NAF) while AT&T refers to the PICC as the Carrier Line Charge (CLC). Sprint refers to the PICC as the Presubscribed Line Charge (PLC).

- Universal Service Fund Charge (USF): The USF is somctimes derisively called the Gore Tax. Vice President Gore championed it and the money collected is designed to help connect rural citizens, schools, libraries and Hospitals to the Internet.

II. Auditing Leased Line Circuits

The second portion of your audit will involve checking the billing for leased line circuits. The following are common types of IXC leased line circuits:

- Dedicated Access Lines (DAL) include T-1s that connect a customer directly to the IXC POP and Point-to-Point Leased Lines. Also includes analog and digital circuits that connect two or more customer locations in different LATAs. AT&T and MCI bill these circuits on invoices that are separate from the main voice invoice. Sprint combines these charges with the main voice usage bill.

- Analog/Digital interLATA point-to-point circuits. This includes all data circuits that directly connect one location to another.

IXCs do not have the physical plant necessary to directly connect into most companies. They have to buy this access from a LEC or from a Competitive Local Exchange Carrier (CLEC). When a customer orders a T1 from their IXC, the IXC in turn orders the portion of the T1 that connects the customer's premises to the IXC POP from the LEC. The LEC becomes a sub-contractor to the IXC. The IXC obtains these circuits at a discount (due to volume and special contracts), marks up the cost of the circuit from the LEC, adds the charges listed below and bills the customer for the T1. If there is trouble on the T1, the customer notifies the IXC and the IXC works with the LEC directly to correct the problem. As far as the LEC is concerned, the IXC is the customer. If a customer tries to call the LEC directly they will be advised to call the IXC. A customer also has the option to direct their IXC to use a CLEC for the local portion of the T1. This is referred to as Customer Non-LEC Access (CNLA).

A customer can also order the T1 to the POP directly from the LEC. This is called Customer Provided Access (CPA). The monthly charge is discounted because you avoid the IXC mark-up. Though CPA makes sense from a financial view, some companies do not use CPA for other reasons. If the T1 has trouble they will have to contact both the LEC and the IXC to see where the trouble lies. They would rather have one point of contact.

Billing Components of an IXC billed Leased Line Circuit

- Access Coordination (AC) Charge – Monthly recurring charge that provides for the design, ordering, installation and testing. The AC also applies to CPA since the IXC will be providing ongoing monitoring of the T1. In some instances, this charge is waived for CPA if the Customer agrees to provide this function internally.

- Central Office Connection (COC) Charge – Monthly recurring charge for connecting the inter-office channel of a dedicated access line.

- Local Access Channel - Monthly recurring charge to connect a customer to the POP. Local channel charges are detailed in the following tariff sections: MCI FCC# 8, AT&T FCC# 11 and Sprint FCC# 8. These charges vary by the NPA/NNX of the customer.

- Inter-Office Channel (IOC) - the long haul portion of a dedicated circuit that connects two or more points of presence in different LATAs. This charge is a monthly recurring charge consisting of a fixed mileage component plus variable component based on the number of miles the circuit travels.

- Special Access Surcharge (S25) - On analog circuits, this charge applies per local channel. For DS1 and DS3 circuits the S25 charge is based on the number of equivalent voice channels (24 for DS1 and 672 for DS3). This charge is waived if the customer notifies the IXC that they are exempt at the time the circuit is ordered or if a customer submits a blanket exemption form to the IXC. The S25 charge varies by IXC. A DS0 circuit will be charged approximately $36.25 per month while a DS1 circuit will be charged approximately $600.00 per month

Non-Recurring (Installation) Charges

These one-time charges appear on monthly invoices for new circuits and changes to existing circuits (moves). Some contracts call for the wavier of these charges. Typically, the charge will be billed and followed up by a credit on the next bill. Many contracts have provisions that carry unused credits over to new contracts.

Figure 6.10 Billing Components of an IXC Circuit

Circuits
1-64K
24-64K = T1 Circuit ≈ 1.544 M
28 T1 = 1 DS 3   ≈ 44.5M

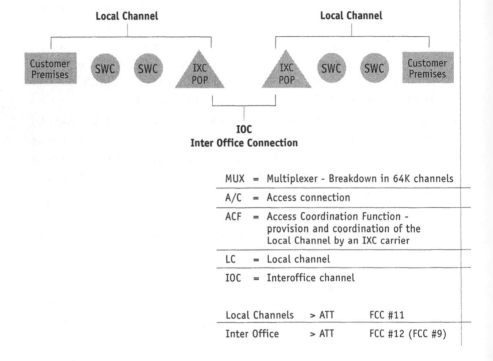

MUX	=	Multiplexer - Breakdown in 64K channels
A/C	=	Access connection
ACF	=	Access Coordination Function - provision and coordination of the Local Channel by an IXC carrier
LC	=	Local channel
IOC	=	Interoffice channel

Local Channels > ATT	FCC #11
Inter Office > ATT	FCC #12 (FCC #9)

Leased Line Pricing Tools

There are a number of leased lines pricing tools that allow auditors to calculate LEC and IXC leased line charges independently. Most are quite easy to use. All you have to do is pick the IXC or LEC that bills the circuit. You will be prompted to enter the NPA/NXX of each location. A schematic of the circuit will appear detailing the circuit and all associated charges. These pricing tools also provide tariff extracts and allow you to check various tariff discounts. Diagram 6.11 is an example of a pricing tool called LATTIS.

Your IXC bill may not itemize the NPA/NXX for each location. In that case you will have to call them and request this list. For MCI and AT&T you can verify the proper cost with the NPA/NXX. See Figures 6.13 and 6.14.

Figure 6.11   **Sample LATTIS Screen**

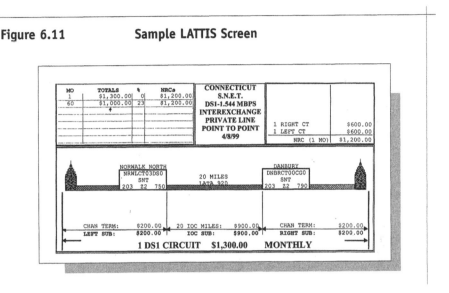

Figure 6.12   **Sample LATTIS Screen**

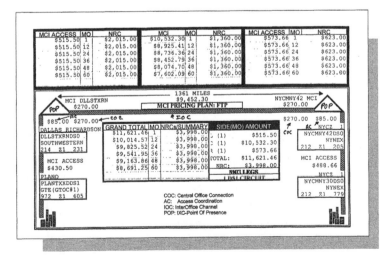

Figure 6.13 MCI NPA/NXX Circuit Billing

NPA NXX	Channelized Access Monthly Price	Channelized Access Install Price	Unchannelized Access Monthly Price	Unchannelized Access Install Price
914 221	$830.43 R	$858.00	$830.43 R	$858.00
914 223	740.40 R	858.00	740.40 R	858.00
914 225	965.30 R	1346.00	965.30 R	1346.00
914 226	830.43 R	858.00	830.43 R	858.00
914 227	830.43 R	858.00	830.43 R	858.00
914 228	965.30 R	1346.00	965.30 R	1346.00
914 229	656.30 R	1346.00	656.30 R	1346.00
914 232	800.50 R	1346.00	800.50 R	1346.00
914 234	759.30 R	1346.00	759.30 R	1346.00
914 235	697.50 R	1346.00	697.50 R	1346.00
914 236	676.90 R	1346.00	676.90 R	1346.00
914 237	718.10 R	1346.00	718.10 R	1346.00
914 238	718.10 R	1346.00	718.10 R	1346.00
914 241	759.30 R	1346.00	759.30 R	1346.00
914 242	759.30 R	1346.00	759.30 R	1346.00
914 243	841.70 R	1346.00	841.70 R	1346.00
914 244	759.30 R	1346.00	759.30 R	1346.00
914 245	821.10 R	1346.00	821.10 R	1346.00
914 246	944.70 R	1346.00	944.70 R	1346.00
914 247	944.70 R	1346.00	944.70 R	1346.00
914 248	841.70 R	1346.00	841.70 R	1346.00
914 249	508.30 I	1346.00	508.30 I	1346.00
914 251	570.91 I	1346.00	570.91 I	1346.00
914 252	1244.03 R	858.00	1244.03 R	858.00
914 253	570.91 I	1346.00	570.91 R	1346.00
914 254	1191.90 R	1346.00	1191.90 R	1346.00
914 255	697.50 R	1346.00	697.50 R	1346.00

NPA NXX	Channelized Access Monthly Price	Install Price	Unchannelized Access Monthly Price	Install Price
914 256	697.50 R	1346.00	697.50 R	1346.00
914 257	779.50 R	1346.00	779.90 R	1346.00
914 258	745.50 I	858.00	745.50 R	858.00
914 259	599.36 I	1346.00	599.36 I	1346.00
914 260	573.90 R	1346.00	573.90 R	1346.00
914 261	367.37 R	1346.00	367.37 R	1346.00
914 262	367.37 R	1346.00	367.37 R	1346.00
914 263	367.37 R	1346.00	367.37 R	1346.00
914 264	367.37 R	1346.00	367.37 R	1346.00
914 265	903.50 R	1346.00	903.50 R	1346.00
914 266	759.30 R	1346.00	759.30 R	1346.00
914 267	738.70 R	1346.00	738.70 R	1346.00
914 268	738.70 R	1346.00	738.70 R	1346.00
914 271	759.30 R	1346.00	759.30 R	1346.00
914 273	676.90 R	1346.00	676.90 R	1346.00
914 276	882.90 R	1346.00	882.90 R	1346.00
914 277	882.90 R	1346.00	882.90 R	1346.00
914 278	944.70 R	1346.00	944.70 R	1346.00
914 279	944.70 R	1346.00	944.70 R	1346.00
914 281	367.37 R	1346.00	367.37 R	1346.00
914 282	367.37 R	1346.00	367.37 R	1346.00
914 283	606.19 I	858.00	606.19 R	858.00
914 284	367.37 R	1346.00	367.37 R	1346.00
914 285	367.37 R	1346.00	367.37 R	1346.00
914 286	367.37 R	1346.00	367.37 R	1346.00
914 287	367.37 R	1346.00	367.37 R	1346.00
914 288	367.37 R	1346.00	367.37 R	1346.00

Figure 6.14 AT&T NPA/XXX Circuit Billing

Rate Table 5.214.B
T1.5 Mbps Local Channels Connected to Switched Services
USOC 1LNV9 - Initial Local Channel
USCO 1LNVT - Additional Local Channel

NPA	NXX	Init'l/Addt'l	Monthly Price	Installation Price
214	200	Init'l	$401.53 I	$1,200.00
	201	Init'l		D
	202	Init'l	$513.19 I	$1,500.00
	203	Init'l	$523.51 R	$1,200.00
	204	Init'l	$408.13 I	$1,500.00
	206	Init'l	$408.13 I	$1,500.00
	207	Init'l	$513.19 I	$1,800.00
	209	Init'l	$408.13 I	$1,500.00
	210	Init'l	$384.23 I	$1,200.00
	212	Init'l	$421.37 R	$1,500.00
	213	Init'l	$362.27 R	$1,200.00
	215	Init'l	$362.27 R	$1,200.00
	216	Init'l	$523.51 R	$1,200.00
	217	Init'l	$523.51 R	$1,200.00
	218	Init'l	$523.51 R	$1,200.00
	219	Init'l	$398.21 I	$1,200.00
	220	Init'l	$432.47 I	$1,200.00
	221	Init'l	$535.13 I	$1,200.00
	222	Init'l	$523.51 I	$1,200.00
	223	Init'l	$523.51 I	$1,200.00
	224	Init'l	$501.14 I	$1,200.00
	225	Init'l	$471.16 R	$1,200.00
	226	Init'l	$523.51 R	$1,200.00

NPA	NXX	Init'l/Addt'l	Monthly Price	Installation Price
214	227	Init'l	$523.51 R	$1,200.00
	228	Init'l	$501.14 I	$1,200.00
	229	Init'l	$410.55 I	$1,500.00
	230	Init'l	$523.51 R	$1,200.00
	231	Init'l	$402.59 I	$1,200.00
	232	Init'l	$421.37 R	$1,500.00
	233	Init'l		D
	234	Init'l	$402.59 I	$1,200.00
	235	Init'l	$402.59 I	$1,200.00
	236	Init'l	$421.37 R	$1,500.00
	237	Init'l	$410.55 I	$1,500.00
	238	Init'l	$250.00	$1,200.00
	239	Init'l	$359.80 I	$1,200.00
	240	Init'l	$421.37 R	$1,500.00
	241	Init'l	$463.57 I	$1,200.00
	243	Init'l	$300.00	$1,200.00
	244	Init'l	$362.27 R	$1,200.00
	245	Init'l	$349.30 I	$1,200.00
	246	Init'l	$375.13 I	$1,200.00
	247	Init'l	$463.57 I	$1,200.00
	248	Init'l	$487.60 I	$1,800.00
	249	Init'l	$408.13 I	$1,800.00
	250	Init'l		D
	251	Init'l	$369.86 I	$1,800.00
		Addt'l	$226.45 I	$260.00
	252	Init'l	$441.05 I	$1,500.00
	253	Init'l	$417.66 I	$1,200.00

NPA	NXX	Init'l/Addt'l	Monthly Price	Installation Price
214	254	Init'l	$349.30 I	$1,200.00
	255	Init'l	$441.05 I	$1,500.00
	256	Init'l	$422.62 I	$1,800.00
		Addt'l	$226.45 I	$260.00
	257	Init'l	$349.30 I	$1,200.00
	258	Init'l	$349.30 I	$1,200.00
	259	Init'l	$365.38 R	$1,200.00
	260	Init'l	$410.55 I	$1,500.00
	261	Init'l	$402.59 I	$1,200.00
	262	Init'l	$410.55 I	$1,500.00
	263	Init'l	$408.59 I	$1,200.00
	264	Init'l	$408.59 I	$1,200.00
	265	Init'l	$479.94 I	$1,200.00
	266	Init'l	$403.90 I	$1,200.00 R
	267	Init'l	$402.59 I	$1,200.00
	268	Init'l	$377.53 I	$1,200.00
	269	Init'l	$349.30 I	$1,200.00
	270	Init'l	$449.00 R	$1,200.00
	271	Init'l	$349.30 I	$1,200.00
	272	Init'l	$529.49 I	$1,800.00
	273	Init'l	$417.66 I	$1,200.00
	274	Init'l	$523.51 R	$1,200.00
	275	Init'l	$402.59 I	$1,200.00
	276	Init'l	$417.66 I	$1,200.00
	277	Init'l	$493.67 I	$1,200.00
	278	Init'l	$417.66 I	$1,200.00

NPA	NXX	Init'l/Addt'l	Monthly Price	Installation Price
214	279	Init'l	$470.66 R	$1,200.00
	280	Init'l	$493.67 I	$1,200.00
	281	Init'l	$349.30 I	$1,200.00
	282	Init'l	$410.55 I	$1,500.00
	283	Init'l	$585.64 I	$1,500.00
	284	Init'l	$523.51 R	$1,200.00
	285	Init'l	$523.51 R	$1,200.00
	286	Init'l	$471.16 R	$1,200.00
	287	Init'l	$602.03 R	$1,200.00
	288	Init'l	$523.51 R	$1,200.00
	289	Init'l	$345.00 I	$1,200.00
	290	Init'l	$395.87 I	$1,200.00
	291	Init'l	$423.59 R	$1,500.00
	292	Init'l	$521.98 R	$1,200.00
	293	Init'l	$423.59 R	$1,500.00
	294	Init'l	$521.98 R	$1,200.00
	296	Init'l	$560.17 I	$1,500.00
	297	Init'l	$410.55 I	$1,500.00
	298	Init'l	$582.44 I	$1,500.00
	299	Init'l	$423.59 R	$1,500.00
	300	Init'l		D
	301	Init'l		D
	302	Init'l	$395.34 I	$1,200.00
	303	Init'l	$401.53 I	$1,200.00
	305	Init'l	$408.13 I	$1,500.00
	308	Init'l	$298.72 R	$1,200.00

NPA	NXX	Init'l/Addt'l	Monthly Price	Installation Price
	309	Init'l	$372.71 I	$1,200.00
	310	Init'l	$358.22 I	$1,200.00
	312	Init'l	$482.68 I	$1,200.00
	314	Init'l	$401.69 I	$1,500.00
	319	Init'l	$448.58 R	$1,200.00
	320	Init'l	$448.58 R	$1,200.00
	321	Init'l	$446.92 R	$1,200.00
	322	Init'l	$408.13 I	$1,800.00
	323	Init'l	$467.64 I	$0.00
	324	Init'l	$402.59 I	$1,200.00
	325	Init'l	$401.53 I	$1,200.00
	327	Init'l	$402.59 I	$1,200.00
	328	Init'l	$402.59 I	$1,200.00
	329	Init'l	$523.51 R	$1,200.00
	330	Init'l	$370.00	$1,200.00
	331	Init'l	$402.59 I	$1,200.00
	332	Init'l	$401.69 I	$1,500.00
	333	Init'l	$482.68 R	$1,200.00
	334	Init'l	$623.52 I	$1,200.00
	335	Init'l	$623.52 I	$1,200.00
	336	Init'l	$408.13 I	$1,500.00
	337	Init'l	$402.59 I	$1,200.00
	338	Init'l	$298.72 R	$1,200.00
	339	Init'l	$402.59 I	$1,200.00
	340	Init'l	$472.79 I	$1,200.00
	341	Init'l	$402.59 I	$1,200.00

NPA	NXX	Init'l/Addt'l	Monthly Price	Installation Price
214	342	Init'l	$480.89 I	$1,200.00
	343	Init'l	$480.89 I	$1,200.00
	344	Init'l	$362.27 R	$1,200.00
	345	Init'l	$402.59 I	$1,200.00
	346	Init'l	$326.56 R	$1,200.00
	347	Init'l	$636.50 R	$1,200.00
	348	Init'l	$480.89 I	$1,200.00
	349	Init'l	$402.59 I	$1,200.00
	350	Init'l	$481.73 I	$1,200.00
	351	Init'l	$481.73 I	$1,200.00
	352	Init'l	$481.73 I	$1,200.00
	353	Init'l	$481.73 I	$1,200.00
	354	Init'l	$421.37 R	$1,500.00
	355	Init'l	$468.28 R	$1,200.00
	356	Init'l	$483.52 I	$1,500.00
	357	Init'l	$481.73 I	$1,200.00
	358	Init'l	$481.73 I	$1,200.00
	359	Init'l	$362.27 R	$1,200.00

When you audit Sprint accounts you need to request both the NPA/NXX and the associated Sprint CLLI Code. Figure 6.15 is a report from Sprint associating the NPA/NXX with the CLLI code. Utilizing the CLLI, we can reference Sprint's tariff (see diagram 6.16) to verify that the proper channel rate is billed.

Figure 6.15 Sprint NPA/XXX– CLLI Code Relationship

Circuit Id	Dedicated Type	Physical Address	State	NPA/NXX Location	NPA/NXX POP	Miles from POP	CLLI Code
172417377	VPN Voice Private Network	1670 Broadway Avenue	CO	303-830	303-292	2	DNVRCOCHCGI
177295487	W9- Toll Free	4350 Weston Parkway	IA	515-226	515-280	1	DESMIAAWDSO
177533500	VPN Voice Private Network	501 Westlake Boulevard	TX	281-366	281-436	11	HSTNTXBUDSO
177819281	VPN Voice Private Network	200 E. Randolph	Il	312-856	312-938	1	CHCGILLRDSO
178663701	VPN Voice Private Network	2815 Indianapolis Road	IN	219-473	219-853	5	WHNGINWTDSO

Request from IXC Information Required to Verify Circuit Component Charges
In this instance Sprint Provided NPA/NNX, Miles from POP, and CLLI Code of Sprint POP

Requested Circuit Information Provided by Sprint

Figure 6.16 Sprint CLLI Circuit Billing

Access and Miscellaneous Services

3. Service Components and Rates (continued)

3.1 Local Access Facilities (continued)

G. Monthly Recurring Line Charge by Local Serving Office (LSO):

LSO	T1 Loop	DAL Loop	Digital Data Service (DDS) 56K Loop[1]	9.6K Loop[2]
PAAUHICO	3765.00	1545.00	1451.00	1401.00
PABGIAXO	783.00	125.00	146.00	145.00
PABLMTXC	2360.00	169.00	225.00	189.00
PACEFLPV	756.00	154.00	246.00	158.00
PACEMSMA	2955.00	321.00	615.00	319.00
PACHCO01	2848.00	96.00	187.00	176.00
PACMCAXF	611.00	119.00	164.00	143.00
PADNOKXA	1403.00	258.00	291.00	267.00
PAGEAZMA	3997.00	133.00	208.00	188.00
PAGENDXA	1021.00	84.00	145.00	134.00
PAGENEXH	2793.00	145.00	231.00	195.00
PAGETXXA	1371.00	181.00	257.00	189.00
PAGVPAXP	1235.00	147.00	219.00	174.00
PAHKFLMA	1399.00	201.00	346.00	208.00
PAIAHICO	2310.00	866.00	788.00	768.00
PALACA11	652.00	131.00	220.00	217.00
PALCKSXA	3154.00	307.00	327.00	284.00
PALNORXX	3090.00	207.00	302.00	230.00
PALOIAXO	691.00	141.00	186.00	147.00
PALOMIXG	1616.00	266.00	335.00	299.00
PALOMNXP	1287.00	123.00	198.00	133.00
PALSWAXX	1357.00	293.00	391.00	369.00

| LSO | T1 Loop | DAL Loop | Digital Data Service (DDS) | |
			56K Loop[1]	9.6K Loop[2]
PALTILPA	583.00	124.00	154.00	175.00
PAMBNJ27	534.00	116.00	263.00	252.00
PAMBNJPM	686.00	124.00	265.00	254.00
PAMPTXPP	1284.00	157.00	208.00	180.00
PANAILXC	1012.00	148.00	308.00	523.00
PANCFLXA	1130.00	314.00	497.00	400.00
PANCNVXF	4561.00	383.00	547.00	392.00
PANLALXA	2451.00	229.00	486.00	294.00
PANLGAMA	612.00	140.00	218.00	144.00
PANMIAXO	1148.00	169.00	204.00	190.00
PANMNEXL	880.00	102.00	132.00	132.00
PANMNYXA	1734.00	314.00	286.00	223.00
PANMOKXA	2455.00	282.00	275.00	246.00
PANRIAXO	1146.00	153.00	204.00	162.00
PAOLINXA	1435.00	170.00	294.00	171.00
PAOLKSPE	885.00	145.00	197.00	171.00
PAOLOKXA	1438.00	218.00	238.00	221.00
PAOLPAPA	613.00	123.00	290.00	239.00
PAONCOXC	2962.00	203.00	295.00	226.00
PARLMDPA	869.00	121.00	340.00	235.00
PARMIDXC	1059.00	186.00	240.00	182.00
PARMMIXJ	1202.00	126.00	173.00	157.00
PARMMOXA	2857.00	294.00	279.00	252.00
PARMOH88	547.00	118.00	155.00	167.00
PARNARMA	903.00	134.00	188.00	163.00
PARSARXA	2221.00	238.00	329.00	206.00

LSO	T1 Loop	DAL Loop	Digital Data Service (DDS) 56K Loop[1]	9.6K Loop[2]
PARSIDXC	4308.00	98.00	216.00	209.00
PARSILXC	800.00	155.00	183.00	178.00
PARSKYMA	810.00	164.00	240.00	171.00
PARMOXA	2417.00	257.00	311.00	226.00

[1] Rates also apply to 64 Kbps speed.

[2] Rates also apply to 2.4, 4.8, and 19.2 Kbps speeds.

Sprint - Local Loop Charges

Sprint F.C.C. Tariff No. 8

A contract may also call for flat rate billing as illustrated in Figure 6.17.

Figure 6.17 Flat Rate Billing

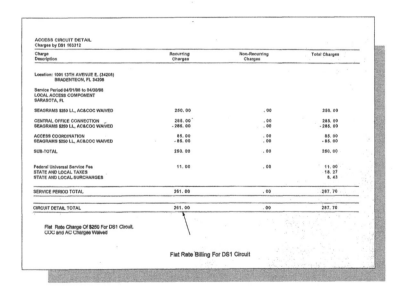

All of the IXCs have special naming conventions for their leased line circuits. It is helpful to obtain these conventions so you can tell what a circuit is being used for, just by checking the Circuit ID. The following is the naming convention used by AT&T.

Figure 6.18 AT&T Circuit ID Naming Convention

AT&T Circuit ID Naming Convention

Example:
DHEC 123456 ATI

Prefix	Service Code	Modifier	Serial Number	Suffix	Company Assigning Circuit Identification	Segment Number
1 2	3 4	5 6	7 8 9 10 11 12	13 14 15	16 17 18 19	20 21 22
	D H	E C	1 2 3 4 5 6		A T I	

Your question involved the "Service Code" portion of the format. Here's a brief list of some of the ones you will see most often.

```
DP  ................  2.4 DDS
DQ  ................  4.8 DDS
DR  ................  9.6 DDS
DW  ................  56K DDS
DS  ................  An old code no longer used for new circuits. Can be any of the above.

DH  ................  T1.5
DN  ................  T45
AR  ................  56K ASDS with DDLC access
AQ  ................  9.6  ASDS with DDLC access
DV  ................  64K ASDS with 64K access
DX  ................  128K ASDS IBR Channel
DZ  ................  256K ASDS IBR Channel
DC  ................  384K ASDS IBR Channel
DE  ................  512K ASDS IBR Channel
DT  ................  768K ASDS IBR Channel

DU  ................  2.048M
DB  ................  ACCUNET Reserved Digital Service (ARDS) 1.5M Access
CY  ................  19.2K DDS
KJ  ................  ISDN B Channel
KZ  ................  ISDN D Channel

FD  ................  Analog Private Line-Data
FT  ................  Foreign Exchange Trunk
FX  ................  Foreign Exchange Line
IT  ................  Inter-Tandem Tie Trunk
OP  ................  Off Premises Extension
OS  ................  Off Premises PBX Station Line
PL  ................  Analog Private Line-Voice
TL  ................  Non Tandem Tie Trunk
TA  ................  Tandem Tie Trunk
TF  ................  Telephoto / Facsimile
WX  ................  800 Service Line
WO  ................  WATS Line (OUT)
WS  ................  WATS Trunk (OUT)
WY  ................  WATS Trunk (2-Way)
WZ  ................  WATS Line (2-Way)
```

Figure 6.19 Sample AT&T DS1 Bill

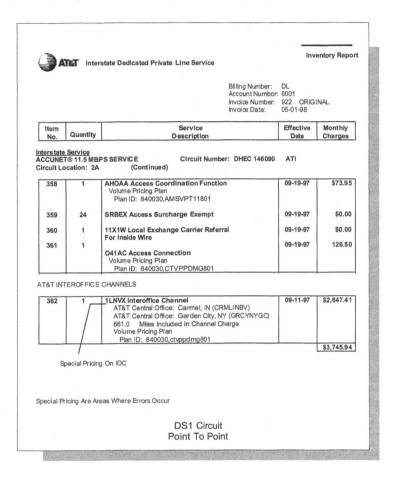

Special Pricing On IOC

Special Pricing Are Areas Where Errors Occur

DS1 Circuit
Point To Point

Auditing Frame Relay bills

Frame Relay service is a packet data service accessible at speeds up to 12.288 MBPS and provides customers with a capability to connect via:

- Permanent Virtual Circuits (PVCs): Associated with dedicated access facilities.
- Switched Virtual Circuits (SWCs): Associated with switched dial-up connections

The following basic charges apply to Frame Relay service:

- Access Coordination (AC) - Recurring Monthly Charge
- Central Office Connection (COC) - does not apply if access facilities are used exclusively for Frame Relay service. A COC charge applies if access facilities are used in conjunction with other services.
- Ports - Frame Relay Ports provide the physical interface to the IXC network for the logical termination of PVCs and SVCs. A PVC is a logical dedicated communication path between two port connections. PVC rates can be billed as a fixed monthly rate or on a usage (metered) rate. Usage is based on the CIR (Committed Information Rate). An SVC is a logical communication path between any two port connections that are set up and taken down dynamically as required.
- Local Loop - Monthly local loop charges apply for dedicated access facilities.

Figure 6.20 illustrates the basic components of a Frame Relay network.

Figure 6.20 Frame Relay Service Components

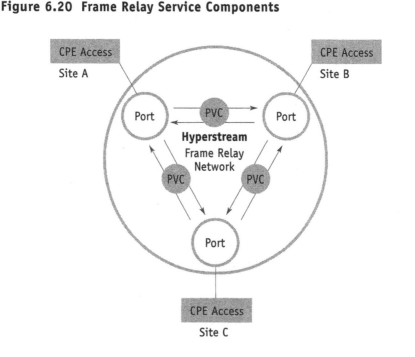

Figures 6.21a and 6.21b are an example of a basic Frame Relay bill.

Figure 6.21a

IXC-Bill Auditing
Access Circuits

Invoice Summary

**** NOTICE ****
Perhaps you did not realize that your last statement was still unpaid. Your payment of $5,002.96
is DUE upon receipt of this statement. If your payment has been mailed, please accept our thanks.

	Previous Balance	
Total Previous Balance		5,002.96
	Payments Thru 02/06/1998	
Total Payments		.00
	Remaining Balance	
Total Remaining Balance		5,002.96
	Current Charges	
Service Charges		
Frame Relay Charges		
Access Charges ←		
Access Coordination		29.75
Local Loop		329.55
Total Access Charges		359.30
MCI Service Charges		
Port Charge ←		250.00
Total MCI Service Charges		250.00
MCI CPE Charges		
Data Service Units		59.00
Total MCI CPE Charges		59.00
Total Frame Relay Charges		668.30
Customer Charges		
Federal Universal Service Fee		
Fed Univ Svc Fee-Prior Month		42.16
Total Federal Universal Service Fee		42.16
Total Customer Charges		42.16
Total Service Charges		710.46
Discounts		
Term Commitment Discount		40.00cr
Total Discounts		40.00cr
Taxes And Surcharges		
State And Local Taxes		3.39
State And Local Surcharges		65.14
Total Taxes And Surcharges		68.53
Total Current Charges		738.99
	Total Due	
Total Due		5,741.95

MCI and all other carriers of interstate communications contribute to federal universal service funds. As a result, MCI
will pass along a subscriber fee to each usage customer. This charge is included on your invoice in the customer level
charges section as a separate line item, on the invoice summary page. The charges included on you February, 1998
invoice, represent January and February charges. Subsequent billing months will typically reflect one month's billing.

Frame Relay Circuit Charges

Figure 6.21b

IXC-Bill Auditing
Access Circuits

Charge Type Summary by Location

	One Time Charges	Usage Charges	Recurring Charges	Credits	Net Total
Frame Relay Charges					
Access Charges					
Access Coordination			29.75		29.75
Total Access Coordination			29.75		29.75
Local Loop			329.55		329.55
Total Local Loop			329.55		329.55
Total Access Charges			359.30		359.30
MCI Service Charges					
Port Charge			250.00		250.00
Total Port Charge			250.00		250.00
Total MCI Service Charges			250.00		250.00
MCI CPE Charges					
Data Service Units			59.00		59.00
Total Data Service Units			59.00		59.00
Total MCI CPE Charges			59.00		59.00
Discount					
HPP Discount ◄————————				-40.00	-40.00
Total HPP Discount				-40.00	-40.00
Total Discount				-40.00	-40.00
Total Frame Relay Charges			668.30	-40.00	628.30
Other Charges					
Customer Charges					
Federal Universal Service Fee					
Fed Univ Svc Fee-Prior Month					42.16
Total Customer Charges					42.16
Total Taxes and Surcharges					68.53
Total Charges			668.30	-40.00	738.99

The Importance of an Audit Test Plan

A thorough audit will examine the IXC contract closely and come up with an Audit Test Plan based on the various discounts and credits promised to a customer. The following is an Audit Test Plan put together for a large customer.

TEST PLAN for Customer XXXX

Lease Line Circuits verify:

Local Loop – obtains 25% discount off tariff

Interoffice Connection (IOC) – obtains 20% discount off tariff

Access Coordination (AC) – obtains 20% discount off tariff

Central Office Connection (COC) – obtains 20% off tariff

Frame Relay Services verify:

20% Local Loop discount off tariff applied

20% discount off tariff is applied to Port charges

20% discount off tariff is applied to PVC usage charges

20% discount off tariff is applied to AC and COC

Toll Free invoices (800 service)

Obtain the following reports from MCI:

- Service Level Summary
- Corporate Charge Invoice Summary
- Summary
- International Minimum Usage Surcharge Detail
- Other Charges/Credits
- Business Unit Summary
- Business Unit Summary Usage Feature Charges
- Service Group Summary Usage
- Service Group Summary Usage Feature Charges
- Monthly Summary of Usage by Toll Free Number
- Busy Days by Toll Free Number
- Busy Days By Service Group
- Call Detail

Verify:

Fixed 8 cents per minute rate

Canadian calls are billed at a fixed 14 cents per minute rate.

Verify international calls receive a 35% discount off the tariff rate on aggregate usage.

Verify that charges for the following advanced routing features are waived:

- Tailored Call Coverage
- Point of Call Routing
- Day of Week Routing
- Time Interval Routing
- Percentage Allocation Routing
- Alternate Routing
- Dialed Number Identification Service
- Network Call Redirect
- Enhanced Call Routing
- ANI

Verify billing for VNET service utilized for outbound switched or dedicated long distance calling.

- Interstate rates – Verify Off-to-Off call are billed at 9 cents. On-to-Off and Off-to-On are billed at 7 cents. On-to-On calls are billed at 5 cents.

- Intrastate rates: Intrastate rates are invoiced at MCI state tariff rates and receive a 20% discount.

- Canada, Mexico, UK calls - Calls made to these countries are billed at a per/ minute rate of 17 cents.

- Installation charges – Verify all installation charges are waived.

A test plan allows the results of your audit to be documented and to be independently audited.

Sample IXC Billing Errors

IXC Case #1

An audit of a large bank found that its custom contract with MCI specifically called for the ANI (Automatic Number Identification) charge of 1 cent per call to be waived. An audit of invoices from 1996 and 1997 determined that the billing of 800 calls were 1 cent higher than what was stated in the contract. At first MCI attributed the difference of 1 cent to rounding. After further review, MCI found that a table update in 1996 caused the ANI to be charged. Another table update fixed the problem in 1997. A credit for 3.1 million dollars was provided to the customer. This error provides a number of lessons. An audit should not only check your current bills but also your back bills for errors. Table updates result in errors that are later corrected. But you may never have been credited for the overcharge. Also, even though IXC billing liability is generally limited to two years, IXCs will go back further under certain conditions.

IXC Case #2

MCI billed a large software company for 8,000 PICC charges. The PICC charge should only be billed on each access line. However, the LECs and the IXCs do not have a satisfactory method in place that allows the LEC to provide the IXC with the exact number of access lines. The MCI bill listed 8000 PICC charges at $2.75 per PIC ($22,000). A CSR was ordered from the LEC. Our audit found only 37 DID trunks (access lines) were billed to this account. The customer had 8,000 DID numbers which are not access lines. DID numbers should not be billed a PICC. The proper PICC charge was determined to be 37 times $2.75 ($101.75) resulting in an overbilling of $21,898.25 per month. A credit of $394,168.50 was calculated back to the inception of the PICC surcharge.

IXC Case #3

A large airline was billed 3% federal tax on all of its usage. As a common carrier, portions of the airline's usage were exempt from federal tax. A credit of $75,000 was provided to the airline by AT&T. Some IXCs will require that you file with the IRS to obtain a refund on taxes collected in error.

IXC Case #4

Federal tax of 3% was charged on calls made from a company on-net location to another company on-net location over a virtual private network. Calls placed over a private network are exempt from federal tax. AT&T credited the account $75,000.00 for federal tax paid in error over the last two years.

IXC Case #5

A large financial services company negotiated a discount of 30% off all international calls with the exception of calls to the UK, Mexico and Canada. These three countries were to receive a 75% discount off the tariff rate. This custom contract was extremely complex and required MCI to manipulate its billing system in a unique fashion. MCI could not provide a line discount of 75% for Canada, the UK and Mexico and a separate line discount of 30% for all other international calls. MCI decided to provide the 30% discount as a line item discount and apply it to all international calls. UK, Canada and Mexico still required an additional discount of 45% (75% - 30% line item discount). MCI went ahead and attempted to hardcode the additional discount of 45% at the call detail level. A typing error by a MCI clerk hardcoded the additional discount as .45 (.45 of 1 per cent) instead of 45%. Because the discount was imbedded into the call detail and was not shown as a separate line item the error was not found for over one year. A credit for $950,000.00 was provided by MCI to this customer.

IXC case #6

This MCI custom contract called for all T1 installation charges to be waived up to 1 million dollars. MCI would bill for the installation charge and then follow up on the next bill with a credit to offset the installation charge. An audit of all the credits found that the credits matched the installation charges. However, the credits did not include the taxes billed on the installation charges. An additional credit of $70,000.00 was provided to the customer.

IXC case #7

This AT&T customer was billed a flat rate of $292 per T1 for 10 T1s for a total of $2920 a month. A close inspection of the contract found that the first 10 T1 charges were supposed to be waived. A credit was issued for $82,000.00 by AT&T and the billing was stopped on a go forward basis.

7

Secrets of
Contract
Negotiation

Telecommunications procurement is not a career for the faint-hearted. Business users have more options today than they did a decade ago, both in terms of technology and suppliers. But unless you can harness these opportunities and use them to your advantage, you may find that you have made commitments you can't meet to buy services that may not meet your needs a year from now and to pay prices that you will soon regret. This chapter is an attempt to level the playing field by explaining the legal framework under which most telecommunications contracts operate, pointing out key terms on which sophisticated large users tend to focus, and offering some pointers regarding negotiation strategy.

Regulatory Background

The Communications Act of 1934 was based on a regulatory regime that was developed in an era in which there was virtually no competition for interstate or international communications services. Although similar regimes adopted by the states for intrastate communications earlier in the century have been substantially rewritten in the last 10-15 years, the basic structure of the federal statute still stands.

Under the 1934 Act and the FCC's rules, basic voice services, dedicated access, private lines and frame relay services must be purchased pursuant to carrier tariffs. (ATM is an unregulated service and is offered only under contract.) The law requires carrier tariffs to include the charges for the service and any regulations affecting those charges. Carriers also enter into contracts with their customers to supplement the tariffs. A recent Supreme Court ruling has raised doubts about how customers can enforce any untariffed terms that contradict the tariff or supplement a carrier obligation set out in the tariff. The cautious customer will not sign up until satisfied with both the customer-specific terms negotiated with the carrier and any applicable portion of the carrier's general tariff. If the carrier refuses to allow the customer to review the customer-specific tariff terms before signing up, the customer should require a contract clause stating that the carrier will tariff all contract terms that must be tariffed in order for them to be fully enforceable by the customer. Any carrier that refuses such a request is reserving the right to nullify the deal.

To complicate the regulatory picture, the FCC issued an order in late 1996 requiring the carriers to withdraw their tariffs for domestic interstate services. The order was immediately challenged by the three largest interexchange carriers, who argue that the agency lacks authority to impose such a requirement. These carriers know full well that the 1934 regime gives them rights vis-à-vis their customers that vendors in unregulated environments can only envy. The order has been stayed by a federal appellate court while the agency considers multiple requests for reconsideration. The FCC is expected to rule on those requests by mid-1999, and the appeal will go forward. It is likely that tariffs will go away, either because the FCC has mandated their elimination or because the carriers will be permitted to withdraw them and their larger customers will demand that they take advantage of this option. In the end, this will mean that the most heavily discounted rates charged by the carriers to their largest customers will no longer be public.

The Long Distance Market for Business Services

The major telecommunications carriers have been offering individually negotiated rates and other terms to business customers for a decade now. AT&T's approach is to segment the market into large users (who buy packages of voice and data services under AT&T's Tariff 12) and smaller users (who buy services individually or in packages under Contract Tariffs). AT&T prefers to put customers into Contract Tariffs, where it negotiates very little other than price and commitment level. AT&T will negotiate other terms and conditions only in Tariff 12, which it makes available only where it risks losing a very substantial customer. MCI, Sprint and other interexchange carriers use the same contract vehicle for all customers, but generally offer their larger customers more favorable terms and conditions.

Before You Start

Know your traffic. This means your current and projected minutes of use by call type (on or off-net), by time of day, by state (for intrastate), by country (for international), by month (if your business is seasonal). It means your average call duration and the nature of all carrier-provided enhanced call features. It means your current costs, both before and after discount. It means the number of private lines and frame relay Ports/PVCs -- and the associated bit rate or CIR, location (NPA/NXX), mileage (where applicable) and current price (before and after discount). It means the rate of network churn.

Without this information, you will be unable to compare vendor bids to one another, evaluate your ability to meet the vendor's proposed minimum revenue and other commitments or determine whether you will be able to take advantage of eligibility requirements for discounts and credits.

A Few Words About Price

Prices for interexchange (that is, non-local) services decrease every year. Some of this decrease is attributable to new technology. For example, fiber optic cable has driven down the costs of transport, and most carriers now offer rates that do not vary with distance (so-called "postalized" rates). Much of the decrease is the result of competition in the market for business services. In recent years, business users have begun to buy from carriers other than the Big Three. Unfortunately, WorldCom's merger with MCI eliminated one such option, but LCI's merger with Qwest has boosted it into the ranks of viable alternatives.

Although everybody knows that prices are declining, many business users are not aware that the carriers' standard prices for most services are rising. In fact, the rates for business services (AT&T's SDN and Megacom, MCI's Vnet and DAL 800, Sprint's VPN and 800 Premiere) have increased 7%-10% per year for the last decade. Real rates are declining because virtually all customers receive some discount off those standard rates, whether under a standard term plan or a negotiated, customer-specific arrangement. As the standard rates have increased, customers signing new contracts have been given larger discounts.

This is not just a matter of historical interest; there's an important lesson here. If you sign a contract with rates that are stated as a percentage discount off the standard tariff rate, it is absolutely certain that the price you actually pay will increase over time. You will continue to receive your stated discount, but increases in the "base" rates will push your prices up. The only possible exceptions to this are rates for international services and dedicated local access.

Finally, users should consider expanding the entities that can take service under the contract, with all usage counting toward the discounts tiers and minimum revenue commitments. You will have to overcome the carrier's natural hostility to what it may view as resale, but the effort may pay off by boosting your discounts and your ability to meet commitments. You will want to make sure that the contract requires the carrier to bill each entity separately. And you will want to be the payor of last resort for those entities that you do not control. For additional price concessions, smaller and mid-sized users should also explore affinity contracts and similar arrangements offered by some carriers to members of particular trade associations.

Contract Terms and Conditions

Following is a list of key issues that large business users typically want to address in contracts for telecommunications services. If any of these are critical to your company, you should know that the carriers are willing to address them. Success in any particular case will depend on two factors: (1) How much does the carrier want to win or hold onto your business? The answer will depend upon the size of the deal and upon whether your business can offer the carrier a degree of legitimacy or a foot-in-the-door in a particular industry. (2) Must the carrier make the requested concessions in order to win or hold onto the business? The signals that a customer gives off are crucial here. A customer that has solicited competitive bids is giving off very different signals than a customer whose chief information officer plays a lot of golf with the carrier's sales vice president.

Terms That Affect Your Costs

Rate Stability -- Carriers will stabilize some or all contract rates, the most common being charges for interstate voice and the port and PVC charges for frame relay and ATM. Although private lines, dedicated access and intrastate and international voice services are often priced at a discount off standard rates, it is possible to secure fixed rates for a dozen or more states and foreign countries, while leaving all other state and country rates to "float" with tariff.

Rate Reviews -- In a world of falling prices, business customers are rightly concerned that rates that are competitive today will be excessive within 12-18 months. The most reliable way to minimize this risk is to maximize your flexibility. Do not commit to a contract term of longer than three years. The longer the term, the longer you must wait for the day when you can renegotiate your rates with maximum bargaining leverage. Do not commit more than 70-80% of your anticipated "spend" for services covered by this contract. The "cushion" of uncommitted business is your life raft. It can provide credibility to your threats to bring in another vendor if the incumbent does not adjust your rates during the contract term to bring them in line with the market. It can protect you from shortfall penalties if your company sells off a subsidiary or otherwise downsizes -- or if you begin to use technology in ways that were not anticipated when the contract was signed (IP telephony, ATM).

If you are signing up for a commitment of over $5 million a year, you may also be able to negotiate some form of rate review provision. The most common version requires the carrier to adjust the rates in your contract to match more favorable rates in the carrier's recent contracts of similar size for other customers. This approach works only if you can verify that your contract is being compared to the most favorable other contracts. If and when the carriers stop publishing the rates for those other contracts in their tariffs – as the Federal Communications Commission has proposed – you will be unable to protect yourself against unfair treatment. There are several alternatives to consider, including the use of a neutral third-party to review and recommend rate adjustments. Even the strongest rate review clause is no substitute for the leverage offered by a low commitment and a contract term of no more than 3 years.

Installation Waivers -- Carriers often waive or offer a fixed credit for installation charges. They also commonly require that any waived charges be repaid for facilities removed in less than some minimum period, usually 12 or 18 months. This means that the carrier recoups the "waived" charge through the monthly charges collected during that period.

Avoiding Hidden Costs -- This is surprisingly difficult to do in contracts for tariffed services. No one in his right mind would enter into a contract in which he commits to spend millions (or even hundreds of thousands) of dollars each year if the vendor told him that additional charges could be imposed without his consent – or even his advance knowledge. But that is precisely what most customers are asked to do when they sign contracts for tariffed services. One major vendor includes references to its tariffs (which can be changed without the customer's consent) in its contracts for unregulated services like private line and frame relay links between non-U.S. locations and ATM service.

These hidden costs find their way to your invoice because the carriers' customer-specific tariffs (except for AT&T's Tariff 12) incorporate the terms of the their standard tariffs. In this way, charges that were never negotiated or even revealed in the carrier's bid may appear on your invoice. (Access Coordination, Central Office Connection and similar fees are common examples.) In addition, when the carrier modifies its standard tariff by, for example, adding a surcharge for Universal Service or increasing the kinds of taxes that may be passed on to customers, those customers who negotiated contracts will find a surprise on their invoices.

Payment Terms -- Under traditional tariff rules, carriers can cut off service for nonpayment after 30 days, but the carriers are sometimes willing to lengthen the period for large business customers. They may also be willing to permit the customer to withhold (or escrow) payment of any charges disputed in good faith.

Billing and Audits -- Larger customers tend to be interested in billing design and functionality, and most vendors are willing to have the contract address billing procedures and format. Also, AT&T now offers business customers a 120-day billing guarantee on switched service. That is, any charges that are not billed within that period are waived. Other vendors are sometimes willing to offer a similar guarantee. Finally, carriers are often willing to permit billing audits up to once a year and more frequently if the customer has genuine concerns about billing accuracy.

Network Optimization – Once a contract is signed, the carrier's sales team's only incentives are to sell you more services, not to assure that you have the leanest network required to meet your needs. To remedy this, some customers require periodic carrier reviews of network configuration and service mix to assure that the package remains as lean as possible.

Terms That Affect Your Contract Obligations

Minimum Commitment -- All carriers require customers to commit to purchase a certain volume of services during the term. As noted above, it is important to keep your commitment to a reasonable level in order to preserve an adequate "cushion". The carrier will resist your efforts to do this, and will tell you that a smaller commitment will mean a higher price. But remember that rates do not depend solely on the size of the commitment, and you can probably find other contracts from the same carrier with lower commitments and comparable or better prices than what the carrier has given you as its "best and final" offer. (There are 20,000+ such filings, and you may need help from a consultant with a good database to find the ammunition you need for your negotiation. However, the increased savings over the term of the deal will likely cover this additional cost.)

Your account team's compensation depends in part on the size of your contract commitments. This arrangement can create an incentive for the team to mislead you into signing up for more than you can deliver. Be sure to verify any information the carrier tells you about your current "run-rate." Does it include services that will not be covered by the new contract? Examples include Internet, X.25, local service and ATM. Does it assume the continued application of the current rates? If the new contract reduces your rates, you will spend less. Does the proposed commitment take into account any anticipated downsizing, including the sale of a subsidiary or operating division? Does it take into account any anticipated technology migration? Examples include the shift from catalog sales supported by toll-free service to Internet commerce and the shift from private line data communications to frame relay or ATM.

You should also try to get flexibility in how the commitments will be satisfied. Here are some items to consider:

- Permit telecom services purchased outside the contract to count toward the commitment.

- Use annual, not monthly commitments; this can be critical if there is any seasonal component to your telecom usage.

- Use an aggregate commitment over the term of the contract. Vendors may also require small annual minimums to ensure a steady cash-flow.

- Include the right to shift some portion of each year's commitment into a subsequent or prior year, or into a work-off period after the contract expires. Such provisions give the carrier the revenue for which it bargained without requiring the customer to pay for service it does not need.

- Minimize the number of sub-commitments for particular services. Carriers often want sub-commitments on particularly lucrative services like international voice.

Adjustments in the Minimum -- All of the carriers offer some form of "business downturn" clause. In some of them, the carrier agrees to reduce the commitment levels if certain events occur. In others, the carrier agrees to talk to you and change the commitments if it wants to. Obviously the latter contract clause has no "teeth", but some carriers (AT&T in particular) tend to give effect to the parties' intent anyway. You will want to make sure that the clause applies to the sale of a subsidiary or business unit, a reconfiguration of your network or a migration to new technology that reduces your needs for the services covered by the contract. And you will want to eliminate any requirement that the clause is restricted to business downturns "beyond the customer's control" or to situations in which the customer cannot meet the commitment "despite its best efforts." Such phrases may void any rights that the clause otherwise guarantees.

Some carriers will also reduce the minimum commitment if the customer discontinues one or more of the services because of persistent performance problems. The standards for such performance failures are negotiable, but they generally involve repeated outages over a fixed period, often 90 days.

Shortfall Charges -- If you fail to meet your minimum volume commitment, the carrier does not receive the benefit of its bargain. For this reason, all carriers impose some sort of shortfall charge. But how much harm does the carrier suffer? If you do not use the service, the carrier avoids certain costs, most notably local access costs -- which represent 40-45% off the cost of providing switched voice service. AT&T, which had consistently required the customer to pay 100% of the shortfall amount, now cuts that penalty in half in new contracts. Other carriers will negotiate these charges if the customer insists. Remember, though, your best protection in this area is a low commitment.

In contracts with sub-commitments, there is typically a shortfall charge for failing to meet both the overall commitment and each individual sub-commitment. Thus, a single shortfall can subject you to two shortfall charges. It is a simple matter to eliminate this double penalty, and the carrier should do it.

One final point about minimum volume commitments. Some contracts state the commitment in net terms, that is, after application of all discounts, while others state the commitment in gross terms. When a carrier uses a gross commitment with a 100% shortfall penalty, the customer can end up paying in shortfall penalties far more than it would have paid if it had used the service.

If the gross commitment is $5M and you spend only $4M (prior to discounts), you pay a 100% shortfall charge of $1M.

The same commitment stated in net terms (i.e., after application of, let's say, a 45% discount) is $2.75M. And the same level of gross purchases is $2.20M when calculated on a net basis. Your 100% shortfall charge would be only $550,000.

Monitoring Conditions -- By law, each customer-specific tariff must be available to any customer who qualifies for its terms. The carriers subvert this requirement, however, by including "monitoring conditions" that only the original customer can meet. Such conditions may include a required percentage of interstate, a maximum percentage of switched-to-switched traffic and/or a minimum call duration. These conditions should be of little concern to you, but only if there is no doubt that you will be able to meet them throughout the term of the contract. These requirements can pose serious problems if, for example, your company increases the efficiency of its call centers by reducing the call holding time or moves a call center to a new location so that previously interstate calls will now be classified as intrastate. To limit this risk, you should limit the number of any monitoring conditions and make sure that the ones you do accept are consistent with your present calling patterns (know your traffic!) and leave plenty of room for change. Finally, make sure that the punishment fits the crime. Thus, for example, if your switched-to-switched calling exceeds the agreed-upon limit, any per-minute surcharge should apply to only those minutes that exceeded the limit.

Exclusivity -- Exclusive arrangements have become increasingly popular among the major carriers in recent years. Some carriers include them in nearly every draft contract offered to their customer. Others require it in return for rate stability or use it to increase the minimum revenue commitments each year (for example, each year's dollar commitment is set at the greater of a specified amount or 90% of the prior year's actual purchases). A user should agree to exclusivity only in exchange for real rate stability, a meaningful rate review process and performance guarantees backed by useful remedies. Remember, once you agree to exclusivity, your incumbent vendor has a monopoly on your business for years to come. If your incumbent carrier's service experiences repeated failures, or your rates are now sitting well above the market price, or you want to migrate to a new technology but the incumbent (knowing that you have no alternatives) is offering above-market rates, or your company acquires a business that uses another carrier -- you have no recourse whatsoever. Exclusivity (whether in its purest form or in the form of an escalating minimum revenue commitment) eliminates all of the bargaining leverage that would otherwise be provided by the 20-30% "cushion" discussed above.

Your Termination Rights

Termination for "Cause" -- It should go without saying that one party may terminate a contract if the other party breaches a material term. Most carrier tariffs do not give the customer this right, although such provisions can be negotiated and included in customer-specific tariffs.

Early Termination Charges -- Customers often want the right to walk away from a contract upon payment of an early termination charge. The carriers accept this, but usually seek to impose a substantial penalty, sometimes as much as 100% of the minimum commitment for the remainder of the term. As discussed above in connection with 100% shortfall charges, these "take-or-pay" clauses are unreasonable and should be reduced to 50% or less. You will also want to resist any requirement that you repay "all credits" received under the contract. Such a broad requirement can include credits against installation charges, outage credits, credits that represent your discounts on intrastate services, and even credits that are refunds for overbilling.

Transitional Support -- Once your contract for tariffed services ends, the rates revert to the carrier's standard rates unless you negotiate another contract with the same carrier. How much time will it take to migrate to a new vendor when this contract expires or if it terminates early? Will the minimum revenue commitments and monitoring conditions permit you to complete the migration in the months before the expiration? (Measurement of revenue commitments and monitoring conditions on an annual basis is clearly more favorable than monthly measurements in this regard.) If not, you may want to negotiate the right to continue to receive the contract pricing for a few months after the expiration without additional commitments. Such transition rights may be even more valuable in the case of early termination, where you may need extra time to prepare for migration to another vendor.

Terms That Affect The Carrier's Performance Obligations

Implementation and Acceptance -- Users contemplating a large-scale reprovisioning of a network typically include a detailed implementation plan as an attachment to the contract. This would certainly be advisable in connection with the migration from one carrier to another or from one service to another. In the past, only large users were able to negotiate remedies for installation delays, but some carriers now include these as standard features in their tariffs and other carriers should be willing to match their competitors in individual cases. For migrations of data networks, carriers have been willing to permit customers to test new service installations and postpone the start of billing until the service passes the agreed-upon tests; the acceptance cycle is typically 2-5 days.

Performance Specifications – Because carrier "fitness standards" typically speak only in terms of "objectives", customers must negotiate in order to get real performance specifications and sanctions for failure to achieve them. Some carriers have begun to include such requirements in their tariffs. For example, AT&T and Qwest have service level guarantees in their tariffs for frame relay service; these guarantees cover items like network availability, round trip delay, packet loss, and mean time to repair.

Outage Credits -- Credits typically offered for outages of switched services are very limited, if they exist at all. Outages of dedicated services typically entitle the customer to nothing more than "run-rate" credits. Thus, if a service costs $300 a month and goes out for 24 hours, you receive a $10.00 credit. Most carriers have been willing to grant higher credits at least in larger contracts and some (as noted above) are starting to include more meaningful outage credits in their standard tariffs. You should know, however, that the carriers do not agree to credits that come anywhere near compensating the customer for the losses it incurs as a result of a carrier outage. In fact, all carrier tariffs preclude carrier liability for so-called "consequential" damages. Finally, do not forget the value of non-monetary remedies for service failures, including a reduction in the minimum commitments for chronic failures or, in the case of critical applications, a requirement that the carrier re-engineer any service elements suffering from chronic failures or even replace them with more reliable technology and absorb the incremental cost of doing so.

Partial Termination Rights -- Most carriers are willing to allow a customer to reduce the minimum commitment to reflect the termination of services that experience chronic outages; the standard for what constitutes a chronic outage is negotiable. The value of this kind of clause is limited, however, in the context of a frame relay or ATM network, where it is not generally considered feasible to mix-and-match vendors.

Carrier Responsibility for Local Access -- While some carriers resist providing any warranties or assurances of performance for local access, AT&T routinely does so. The issue is negotiable with other carriers, and is important if you are looking for end-to-end performance guarantees. All carriers will permit customers to purchase dedicated access directly, and large users sometimes realize substantial savings from contracts with alternative providers. You should know, however, that the interexchange carriers may try to impose certain connection charges for customer-provided access that will wipe out any savings. These charges (access coordination, central office connection, entrance facility and others) can be negotiated.

Terms Relating To Service Support For Larger Customers

Larger customers typically require more in the way of customized support services and other terms. These may include the following:

- The right to have input on the choice of personnel assigned to the customer's account.

- Reasonable restrictions on its employees' rights of access to sensitive customer locations.

- Extensive documentation (technical and training manuals, etc.) relating to their services.

- Information and advice regarding the compatibility of the carrier's services with equipment the customer is considering buying.

- Periodic reports relating to satisfaction of all commitments and monitoring conditions, network usage, circuit inventory, etc.

- Access to "on-line" network management -- order entry and tracking; trouble entry and tracking; network monitoring.

Terms Addressing Important Risk Allocation Issues

Tariff Changes – Carriers carefully guard their right to modify their tariffs at any time. Although there is legal precedent on which you may rely to challenge any effort to change the rates and other terms contained in your customer-specific tariff without your consent, the law offers little comfort with respect to changes in the general portions of the carrier's tariff, even if they are clearly applicable to you. For this reason, the carriers incorporate the terms of their general tariffs into each customer-specific tariff. (AT&T's Tariff 12 departs from this model and is, therefore, a much more desirable contract vehicle for customers.)

Intellectual Property Indemnification -- This issue is gaining importance with the introduction of new enhanced features in the carriers' networks and in equipment used by the customer. For example, AT&T has a long-standing patent-infringement suit against its competitors relating to their use of technology involved in providing toll-free service. MCI has now brought suit against AT&T for violation of patents involving interactive call processing technology. (MCI did not develop the technology but bought from the patent-holder the exclusive right to enforce the intellectual property rights against AT&T.) Although customers have not been named in these suits, their contracts have been sought as evidence.

There are two risks here. One is that the customer will be inconvenienced by a patent, copyright, trade secret or similar lawsuit between its vendor and others. The other is that the customer may itself be sued in such a case. You should require your carrier to agree to defend you and pay all damage awards and other costs if you are sued because of your use of the carrier's service. The costs incurred by the carrier under such an indemnification should not count against any limitation on the carrier's liability that may be included in the contract or the carrier's tariff.

Year 2000 Protections – There are several areas in which a carrier's Y2K problems can adversely affect the customer:

- Transmission networks can fail if the software in the carrier's switch misreads the "00" in the date field and shuts down because it believes the switch has not been maintained since 1900.

- Carrier billing systems can fail or generate errors on customer invoices.

- Administrative support systems involved in ordering, provisioning and network management and reports can fail and disrupt service.

Many carriers report that their Y2K remediation programs are on schedule and that they anticipate being fully compliant by mid-year, and most are now providing detailed information about their efforts on their Web pages. As recently as a year ago, carriers offered representations and warranties that their services and systems would be compliant by June 1999. Some users negotiated remedies (including the right to terminate) if the services were not compliant by that date. But as the millennium approaches, these same carriers are suddenly reluctant to offer such legal protections.

You should certainly try to negotiate Y2K warranties. Failing that, you might look for protections in the areas listed below:

- Credits and/or termination rights for sub-standard network availability

- The right to withhold payments if invoices are inaccurate, and a waiver of charges that are more than 120 days after they were incurred

- Credits and/or termination rights for delayed installations.

In addition, you will want the carrier to agree that it will not claim that a Y2K failure in the systems it uses to provide the service is an event "beyond its control," even if the failure was caused by software supplied by a third party (for example, a Lucent or Nortel switch). Carrier tariffs all excuse the carrier's failure to perform if it is the result of an "Act of God" or other causes beyond the carrier's control. So-called "force majeure" provisions are reasonable, but should not permit the carrier to escape responsibility for the technology that it incorporates into its systems. And, you will want to make sure that you have a right to cancel any service that suffers from Y2K problems and that your minimum revenue commitments are reduced accordingly. In the end, however, the paramount concern with Y2K is ensuring continuity of service – not outage credits or termination rights. Even with the best contract clause, you should not ignore the importance of contingency planning.

Fraud Risks -- Under carrier tariffs, the customer must pay for all calls originating on its network. This includes calls placed by individuals who "hack" your PBX. In nearly all circumstances, a customer has no knowledge of such hacking until it receives the carrier's invoice 10, 20, or 30 days later. By then, the damage has been done, and it can be in the tens of thousands of dollars or more. Carriers are in a position to notice unusual spikes in your traffic or changes in your calling patterns as they happen and should, therefore, be required to notify you immediately of potential problems. But because their tariffs place the entire risk of PBX fraud on the customer, the carriers have no incentive to act. (Compare this with calling card fraud, where the carrier bears the risk and, therefore, acts promptly to shut down fraud as it is happening.) The Federal Communications Commission is considering whether to shift at least some of the risk of PBX fraud to the carriers, but don't hold your breath -- the proceeding has been pending since 1993. All of the major carriers have some kind of "fraud protection" program, but be sure to read the fine print. The Sprint and AT&T programs are generally well-liked by customers, but the MCI program has harsh restrictions (your remedy is limited to a once-in-a-lifetime credit of $30,000).

Strategic Advice

The following points offer critically important advice on negotiation strategy and summarize the key items discussed above:

- Do not convey to the carrier that you have already decided to sign the contract before the negotiations are over. Once you send that message, there is no reason on earth why the carrier should agree to anything you request.

- You are sending the wrong message if you start placing orders before the contract is signed, if you tell the carrier that you have to sign the contract by a certain date, or if you express distaste for the negotiation process. It is okay to have deadlines or to express impatience with the process, but you should control the process so that these work for -- not against -- you.

- Make sure that you and your management are sending the same signals to the vendor. Deals invariably go sour when someone in authority undermines the negotiating team with an ill-timed wink and nudge.

- Stabilize as many of the rates as you can, so that they do not increase as the carrier's standard rates rise over the term.

- Require your account team to list all of the charges that will appear on your invoices. While you may not succeed in locking these in, this approach may increase your ability to hold their feet to the fire in future negotiations.

- Make your contract term no more than three years and keep your minimum revenue commitment low. Even if it costs you some savings at the start, a 20-30% "cushion" of uncommitted traffic will save more money in the long run in the form of meaningful rate reviews and avoided shortfall penalties.

- Verify any information the carrier tells you when determining what your commitment should be.

- Make sure that shortfall and early termination charges provide reasonable compensation for the carrier's losses.

- Require your account team to provide a copy of the filed contract tariff within 2-3 weeks after the contract is signed. Check the tariff carefully and bring to the carrier's attention any missing terms or items that are not consistent with your contract.

- Remember, promises included in a contract or a "side letter" are not legally enforceable if they contradict or supplement any carrier obligation addressed in the carrier's tariff. This is true even if the contract or side letter is signed in blood by the Chairman of the Board.

8

Utilizing the Power
of the Internet
to Audit Your Bills

The Internet has become so much a part of our everyday life that those among us who resist using it risk being left behind. In addition to being a source of entertainment, it has become an invaluable tool for researchers, consumers, and businesses alike.

Within the context of this book, the Internet is a great source for information about equipment vendors, tariffs, regulators, local, long distance, and other carrier services. You can also find out about telecommunications trade associations and industry publications, consultants and research organizations on-line.

Once you've begun to understand your CSR and how USOCs and other service order mechanics affect billing, you're ready to use the Internet to maximize your telecommunications assets. The amount of relevant information found in cyberspace is truly mind-boggling and at times can be confusing. This chapter is devoted to showing you how to use the Internet efficiently to ensure your telecommunications environment remains in top condition.

Getting Started – Search Engines

A good way to find information on the Internet is to use search engines. There are several well-known search engines including AltaVista, Yahoo, Excite, Lycos, and Infoseek.

Search engines let you perform keyword searches. When you type in a word (subject, company name, person, etc.) you are quickly presented with a list of relevant web sites and/or sources associated with your keyword.

One problem with keyword searches however, is that you often get more results than you bargained for. Typing in a keyword like "telecommunications" can yield thousands of results. Many are totally useless and unrelated to your research.

Use search engines like AltaVista for very narrowly defined specific searches. Try guides like Hotbots and Yahoo to conduct broad sweeps. Unfortunately, there is no search engine that is all-inclusive. Each search engine specializes in its own corner of the Internet. They each contain millions of home pages, and each is indexed differently. That's why search results can vary so much.

Use what researchers call Boolean Operators like "AND", "OR", "XOR", "NOR" and "NOT" to narrow your search and increase your chances of getting meaningful hits. Typing 'Sprint' AND 'long distance' will provide hits about the long distance carrier, not Track and Field events. Also, use exact and unique search terms. Be descriptive. Whenever possible, add keywords or industry-specific phrases. If you want to learn about a callback provider identified as "XYZ Company Incorporated", don't just type in "XYZ Company." Type in "XYZ Company Incorporated" + 'Callback Operators'.

It's also essential to avoid keyword spelling errors. Spelling errors and typos can frustrate attempts to link up with desired websites.

Search Engine Examples

The following examples show the homepages of some of the more popular search engines. Note the 'search' fields and how each directory lets you define how and what you're looking for.

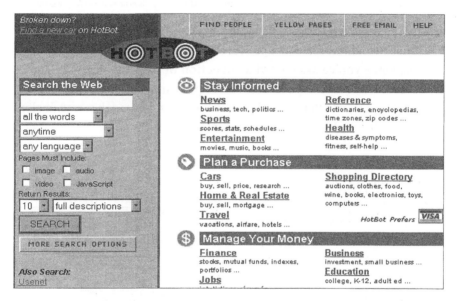

Hotbot   www.hotbot.com

If you're looking for general information, try Hotbot, one of the world's largest Web search engines. It features general information on hundreds of topics grouped into 16,800 categories. Use Hotbot's search engine to look up what you want and then use its pull-down menus to refine the topic even further. Subject menus guide you through categories like Business and Finance or Computers and Internet.

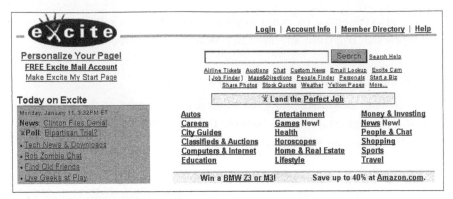

Excite   www.excite.com

Excite is suited for concept searches rather than narrowly defined queries. Once you've assembled a collection of possible destinations, it offers tools to narrow your search. It presents you with ten search results at a time, which it ranks by relevance to your search terms. It also allows you to display your hits by summary or title.

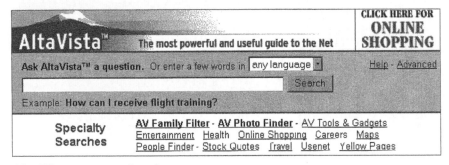

AltaVista   www.altavista.com

AltaVista is considered one of the best search engines among Internet enthusiasts. It provides a fast search engine boasting an index of over 125,000,000 unique World Wide Web pages. Simply enter your query and bracket it with quotes. After you review the results, click on the "Refine" button to get a more exact match.

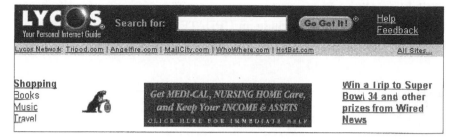

Lycos   www.lycos.com

Lycos is a leading search engine and directory site visited monthly by 15 million users in the U.S. alone. You'll also find lots of helpful links to free software here, and information on a broad range of topics like careers, education and health.

Auditing and Tariff-Related Websites

Tariff information is available on-line as a fee-based service from various companies. Some tariffs are available on-line free of charge from the carrier directly or from regulatory agencies.

Fee based tariff services:

Tele-Tech Services   www.tariffs.com

Telecommunications consultants and auditors will love this site. Tele-Tech will furnish you with comprehensive tariff and rate databases all available on-line. You can also request tariff information sent in print form. Other useful links include an UPDATES e-zine (magazine on-line) on regulatory issues.

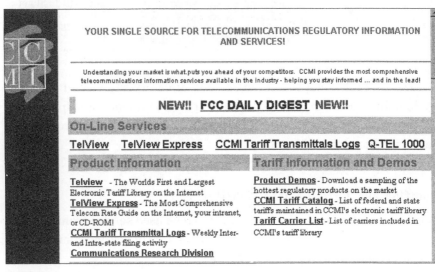

Telview   www.telview.com

This is a great site for anyone needing quick access to telephone tariff and rate information. Telview, produced by the Center for Communications Management Information (CCMI), features the world's first and largest electronic tariff library. With a click of the mouse, subscribers can get instant access to the latest tariffs of most of the country's carriers. Communications managers will also like Telview Express, an interactive telecom rate guide you can run either on the Internet, a corporate Intranet, or on CD-ROM. Check out Telescope, CCMI's monthly newsletter on regulatory trends and issues. Also featured are product demos to test drive the latest tariff software products on the market.

Products & Services

Industry Events

WORLDLYNX Newsletter

Job Opportunities

LYNX Links Page

Lynx Technologies   www.lynxtech.com

LYNX Technologies collects international telecom rate and service data from numerous countries around the world. The LYNX Global Telecom Database provides the most comprehensive coverage available for national and international telecom pricing for over 350 carriers in more than 200 countries. You can find state-of-the-market country profiles and news on upcoming telecom events. And the globetrotter can access a huge library of over 220 international telecom hyperlinks.

...Home of the World's First Internet Tariff Library!

Announcing the addition of a new set of tariff menu options!

Valucom has just added a new carrier based set of menus to the TariffNet site. The carrier-based menus support ALL of the access methods (PDF, TIF, and Text). Our original menu is now called 'By Category.' If you have any questions or comments, please contact your salesperson or customer service.

Main Menu

Welcome to tariffnet.com ... Home of the World's First Internet Tariff Library!

This site is dedicated to providing telecommunications tariffs to our subscribers. Tariffnet subscribers are only a few mouse-clicks away from **complete U.S., Canadian, European, and Pacific Rim** tariffs. Each of the tariffs are updated hourly, as changes are received at Valucom, Inc. You may never have to file a tariff again!

If you have a large number of users requiring access to electronic tariffs or if your connection to the Internet is an issue, Valucom also offers electronic tariffs via your corporate intranet. If you would like

Tariffnet.com   www.tariffnet.com

Tariffnet.com lets you access current tariff and rate information 24-hours a day, seven days a week. Using only a standard web browser, the Valucom Inc., service provides a complete telecommunications tariff library accessible via the Internet. Their electronic telecommunications library contains an exact digitized replica of all pages of the important tariffs that have been filed at state public utility commissions as well as the telecommunications tariffs on file at the FCC. Valucom also offers Canadian tariffs along with tariffs from over 80 major countries around the world.

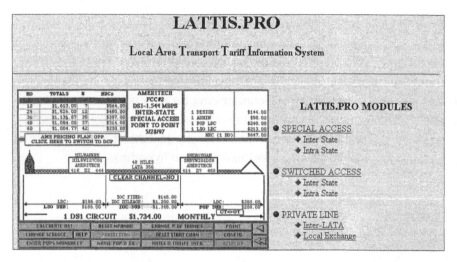

LATTIS.PRO   www.triquad.com/lattis.html

For those doing telephone bill auditing, this fee-based site offers tools to make the job easier. This no-nonsense site is produced by Tri-Quad Enterprises who's creative, intuitive graphical pricing and rate databases are used extensively by large carriers and large business users. Whether it's switched or special access, interLATA or intraLATA rates, LATTIS delivers current and accurate data right to your PC. Their database contains information on circuits covering the entire United States. LATTIS can price hundreds of access circuits at once and can handle all types of network configurations, such as hubbing, multi-point and point-to-point. In addition, LATTIS will show the circuit graphically on your PC screen, breaking down all circuit elements.

Free on-line tariff information:

US West http://tariffs.uswest.com

US West has its FCC tariffs and state tariffs for Arizona, Colorado, Idaho, Minnesota, Montana, Nebraska, New Mexico, North Dakota, Oregon, South Dakota, Utah, Washington and Wyoming on-line.

Bell Canada www.bell.ca

Bell Canada has its tariffs on-line.

Due to increased flexibility being given to the ILECs and AT&T, the FCC is encouraging all carriers to make their tariffs available on-line. Before you pay a tariff service for access to a particular carrier's tariff check their web site and the state PUC's web site to see if the tariff has been made available on-line for viewing and downloading without charge. At the end of this chapter we list all of the major carrier's web sites and all of the state PUC's web sites.

Consultants and Consultant Liaison Programs

Major telecom suppliers have recently been offering liaison programs to enhance their relationships with consultants. They figure the more you know about them, the more likely you may want to conduct business with them in the future. Knowledge is power, especially when negotiating a vendor contract. Consultant Liaison Programs (CLP) offer valuable competitive information and support services that enable consultants to negotiate advantageous contracts for their clients.

When you register for a CLP, you'll gain access to the latest rate information, promotions, collateral material, press releases and product announcements. Most of these programs are now available on the Web. If you are a telecom consultant, auditor or administrator, it is highly recommended you register with one or more of the on-line consultant liaison programs listed below. It might take a few minutes to fill out a profile form, but it is well worth it. They're free of charge and offer valuable information that is just a mouse click away.

STC Overview
Membership
Vendor Advisor Council
STC Conferences
STC Newsletter
Contact STC

Experience
Knowledge
Resources
Commitment
Ethics
Objectivity
Professionalism

STC

Society of Telecommunications Consultants

The Society of Telecommunications Consultants *(STC)* is an international organization of voice and data communications professionals who serve clients in business, industry, service organizations, and government. STC members adhere to strict professional standards and a rigorous code of ethics.

STC has established a set of operating procedures which enhance the professionalism and credibility of its members. STC is dedicated to improving the competence of its members so that they may serve clients with reliability, integrity and the highest level of professionalism.

The Society of Telecommunications Consultants
www.stcconsultants.org

This international organization is comprised of voice and data communications professionals. STC is dedicated to improving the competence of its members so that they may serve clients with reliability, integrity and the highest level of professionalism. It is a major voice in regulatory, legislative and commercial issues. The Society of Telecommunications Consultants also sponsors conferences and publishes the STC Newsletter

Sprint Consultant Liaison Program   www.sprint.com/consult

Sprint provides a plethora of current information to make product or service recommendations easier. You'll get the Sprint Source Newsletter

---

designed exclusively for consultants. In addition, you'll get information about communications solutions for any size business, along with the lowdown on Sprint-sponsored seminars and teleconferences discussing new products and services. Registered members of Sprint's CLP gain access to a members-only section where they can find up-to-date rate information.

Bell Atlantic CLP   www.bell-atl.com/largebiz/lbs/clp_how.htm

Bell Atlantic's CLP offers a one-stop shop for consultants with a CLP e-zine and consultant handbook with ready-to-use materials and contact information. On their Web site, members get quick access to information on products, services, news on company-sponsored training and maintenance programs, press releases and a calendar of upcoming trade shows and industry events.

Fuld Intelligence Index   www.fuld.com

This site offers over 300 intelligence links grouped by industry. Fuld offers tips on devising a competitive intelligence strategy. Use the site's Strategic Intelligence Organizer, which prompts you to ask the right questions when doing your research. Test your investigative ability by taking the Corporate Evaluation Questionnaire to evaluate your ability to handle and apply vital intelligence.

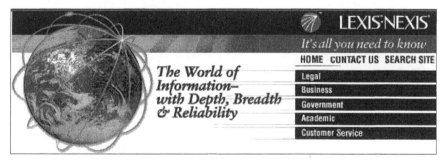

Lexis-Nexis   www.lexis-nexis.com

Lexis-Nexis is another source for business intelligence. Here you can find out if your competitor is embroiled in litigation or has a lien against them in any state. This site offers the availability of over one billion business-related documents with millions more added to their database weekly.

SOCIETY OF
COMPETITIVE INTELLIGENCE
PROFESSIONALS

● Chapters ● Experts ● Contact Info ● CI Forums ● Links ● FAQ

News and Notes for the Week of January 4, 1998

The New SCIP.ORG
Our new Web site is ready, it just has very little content. We are
transferring information now and adding new stuff. By April, we'll have a
resource guide online, our membership directory available, and many more
exciting new features.

Members: Renew Now for 1999
If you joined the Society before July 1, 1998, it's time to renew your
membership (you may have already received a renewal notice). Renew
today using our secure online form.

Society of Competitive Intelligence Professional www.scip.org

This site, produced by a prominent trade organization of competitive
intelligence experts, will help you stay on top of upcoming CI events,
and hot security-related topics. Browse through CI surveys and articles
on intelligence-gathering strategies. Take part in an on-line CI forum in
one of the site's many chat rooms.

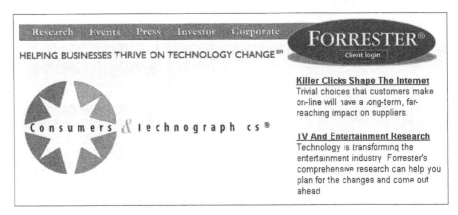

Research   Events   Press   Investor   Corporate

FORRESTER®
Client login

HELPING BUSINESSES THRIVE ON TECHNOLOGY CHANGE[SM]

Consumers & Technographics®

Killer Clicks Shape The Internet
Trivial choices that customers make
on-line will have a long-term, far-
reaching impact on suppliers.

TV And Entertainment Research
Technology is transforming the
entertainment industry. Forrester's
comprehensive research can help you
plan for the changes and come out
ahead

Forrester Research www.forrester.com

Forrester Research is a well-respected authority on high technology
market trends. They offer a variety of retainer programs to access
reports and white papers on emerging technologies. You can register as
a guest to their homepage to tap into some of these outstanding

resources free of charge. Search through their archives to locate reports, technology briefs, and research findings in a broad range of areas like e-commerce, cable modems, and educational software. Registered members receive weekly research updates via e-mail.

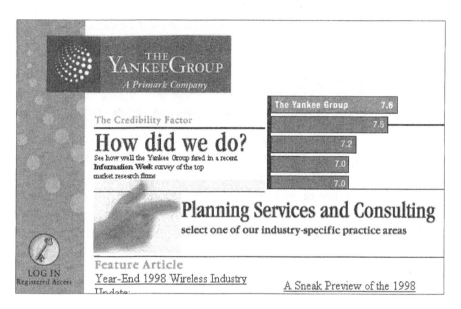

The Yankee Group   www.yankeegroup.com

The Yankee Group is a leading authority in telecom research. The company offers planning and consulting services, as well as top-flight reports on high tech issues impacting the marketplace. Their site is simple in design, and easy to navigate. Once you register, you can gain access to expert commentary and advice, a calendar of industry events, and a searchable archive of executive summaries for all Yankee Group publications. You can also order reports on-line.

STAT – USA   www.stat-usa.gov

The United States Department of Commerce offers a site to help your business grow. STAT-USA features a compilation of trade resources to help you capitalize on global business opportunities. Some highlights include market and company specific research reports, an export yellow pages, on-line newsletters, international trade statistics, and an interactive chart containing up-to-the-minute exchange rates.

USTA's Telecom Policy   www.telecompolicy.net

This site, formerly known as Call Them On It (www.callthemonit .com), is sponsored by the United States Telephone Association (USTA). It's full of policy updates, ILEC activities and their lobbying efforts. Member companies are listed in alphabetical order, and you can follow hotlinks to their websites.

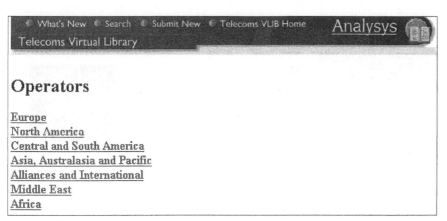

Operators

Europe
North America
Central and South America
Asia, Australasia and Pacific
Alliances and International
Middle East
Africa

Telecom Operators   www.analysys.co.uk/vlib/operator.htm

This impressive site provides an exhaustive listing of just about every carrier in the world. The site is indexed by continent and country to make it easy to find the carrier you want.

On-Line Publications

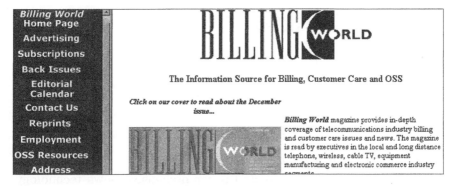

Billing World Magazine   www.billingworld.com

Billing World Magazine (published monthly by Telestrategies) provides in-depth coverage of billing and customer care issues not readily found elsewhere. It's read by top telecom executives, cable TV and utility professionals. The homepage offers impressive links to dozens of Operational Support Systems (OSS) vendors. You can also locate hard-to-find information about billing topics, along with good links to jobs.

Welcome to Phillips TelecomWeb, the leading source of business news, market research, and competitive analysis on the telecommunications industry. **Register** now for a FREE trial! Also, visit our new **Career Center**

Winner, Standard of Excellence Award

Sign up for **TelecomWeb Direct**, a free weekly e-mail service containing the latest additions to TelecomWeb.

Telecom Web   www.telecomweb.com

Phillips Publishing runs the telecomweb.com site. It provides informative company profiles, news, useful industry directories, research reports, and links to trade associations, public utilities commissions and PTTs.

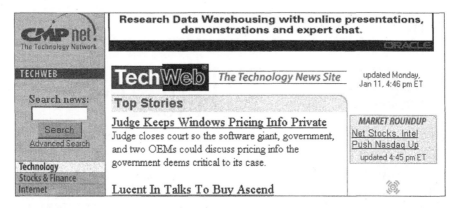

TechWeb   www.techweb.com

CMP Publications has created a comprehensive news technology site that places a wealth of industry information at your fingertips. TechWeb provides product reviews, shareware, news flashes, information about trade shows and a technology encyclopedia. What you're likely to find useful is a search engine that lets you peruse the archives of CMP's seventeen magazines, including Computer Reseller News, Information Week and Internet Week dating back to 1994 with free, full-text retrieval of any article.

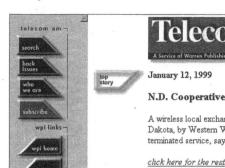

January 12, 1999

N.D. Cooperative Yanks Wireless Provider over Competition

A wireless local exchange carrier (LEC) service launched last week in Regent, North Dakota, by Western Wireless ended abruptly January 11 when the incumbent terminated service, saying Western wasn't authorized to provide LEC services.

click here for the rest of the story ...

Telecom AM   www.telecommunications.com/am/

Telecom AM keeps its ear to the ground on current events in the telecom industry. This fee-based site offers the top story of the day, including federal, business, and mobile news. The site is a service of Warren Publishing, Inc. (WPI), the Washington, D.C.-based publisher of Communications Daily and a wide variety of other newsletters. A one-year subscription to stay plugged in runs about $589. You can take a free trial for two weeks. Registered users can access a searchable archive of daily content dating back to 1996. The site also provides valuable industry links to executive summaries, informative newsletters, and market research reports.

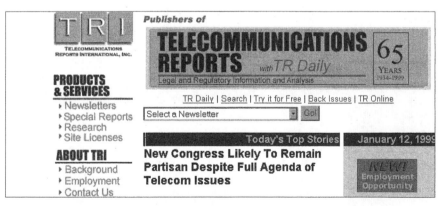

Telecommunications Reports   www.brp.com

This site, sponsored by the publishers of Telecommunications Reports International, helps you gain a competitive edge with timely news updates. You can access 15 telecom newsletters covering a broad range of topics in telecom and multimedia. Click through to special reports, and news on upcoming industry conferences.

Discount Long Distance Digest   www.thedigest.com

Vantek Communications provides a free Web-zine (on-line magazine) on long distance industry news of special interest to long distance carriers, resellers, agents and consumers. Updated regularly, this site offers a monthly wrap-up of major news in long distance. It's also packed with practical information on how to break into the resell market, find the best long-distance sites on the Web, along with what long distance carriers to avoid with their great Hall of Shame section. You can browse through issues dating back to 1993.

FT Media and Telecoms   www.ftmedia.com

You'll find a global perspective of the telecom industry at this site, published by Financial Times, one of the world's leading newspapers for international business. FT Media and Telecoms features a great selection of industry news coverage supplemented by an in-depth assessment of new technologies and market trends. You'll also gain access into an extensive lineup of newsletters, magazines, management reports, yearbooks, and directories covering every aspect of the media and telecommunications industries.

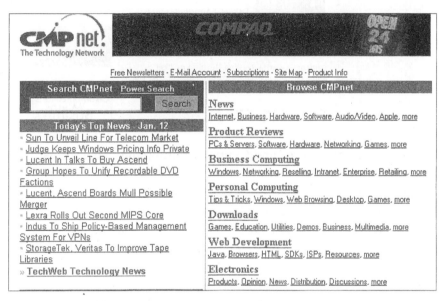

CMPNet   www.cmpnet.com

This is a great resource for anyone who builds, sells, or uses telecommunications and computer technology. This CMP-sponsored site provides a broad array of high tech resources. It covers a full range of technologies organized by topics like browser systems, hardware, and multimedia software. Dozens of hyperlinks to CMP trade publications like Computer Reseller News and Electronic Buyer's News are featured, along with top high tech news of the day and free industry newsletters.

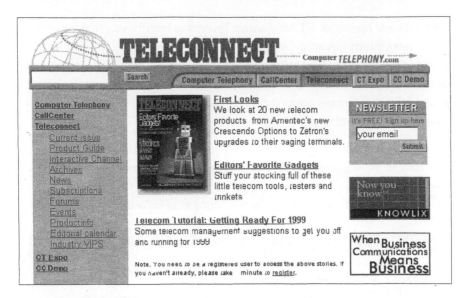

Teleconnect, Call Center and Computer Telephony Magazines
www.teleconnect.com

These three trade magazines are published by the huge, U.K.-based, Miller Freeman, Inc. Teleconnect focuses on how to buy and use key systems, PBXs, voicemail, automated attendants, LANs/WANs, wiring products, call accounting systems, and more. Call Center Magazine is aimed at call center managers and technology buyers and covers the latest ACDs and other call center products and issues. Computer Telephony looks at the convergence of PCs, LANs, and phone systems. It also covers speech recognition, application generator software, fault resilient computers, and more. Miller Freeman also sponsors various trade shows throughout the year, and offers links to information about their shows at this site. Registered users can search through back issues of all the magazines for desired articles and information.

InfoSpace   www.infospace.com

If you only have a phone number or address and need a name, you can use InfoSpace and its on-line Reverse Directory to find it. InfoSpace provides an easy-to-use electronic Reverse Directory. Enter an address and you'll get a name and phone number. You can also search by and for e-mail addresses. This expansive site also lets you search public records and perform background checks. Some of the information is free.

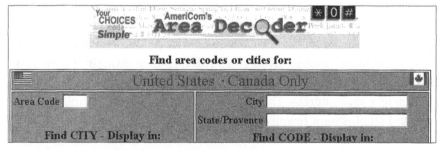

AmeriCom Long Distance Area Decoder
www.decoder.americom.com

If you're looking for the correct area code, check out this site. Americom Inc., a long distance and callback provider, offers an excellent search engine that finds updated area codes and country codes. Simply type in the city and you soon have your information. And if you know the area code but don't know where you're calling, you can search by area code and the site will deliver the city and state or province. For international calling, you type in the city and the Area Decoder will supply you with the city and country code.

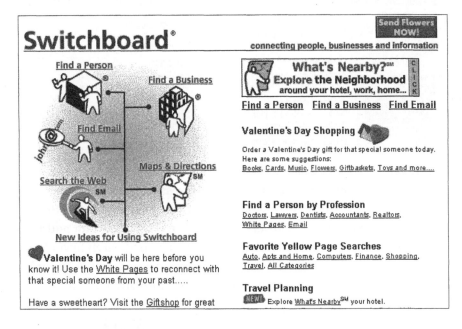

Switchboard   www.switchboard.com

Switchboard provides access to over 106 million residential and 11 million business listings. Even if you don't know exactly where your party is located, you can search by area or near where you think they are or even by e-mail address. There are excellent links to AT&T toll-free listings and employment resources are also featured.

Telephone Directories on the Web   www.contractjobs.com/tel

This site is huge. If you can't find your listing here, it may not exist. You'll find on-line directories spanning the world, sorted by continent and country. You can even explore fax directories, yellow pages, toll-free directories, and e-mail directories furnished for every geographic region.

Telecommunications Organizations And Trade Shows

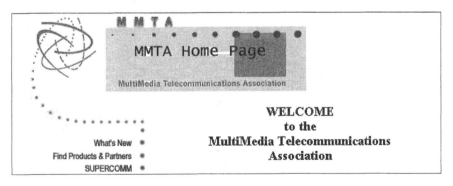

MultiMedia Telecommunications Association   www.mmta.org

Founded in 1970 as the North American Telecommunications Association (NATA), this group is now called the MultiMedia Telecommunications Association (MMTA). Its members are leading equipment manufacturers, value-added resellers, interconnects, developers and consultants. MMTA also publishes white papers, case studies and offers speakers bureaus, training workshops, and market education programs.

United States Telephone Association (USTA) www.usta.org

The United States Telephone Association (USTA) is the leading association for the local exchange carrier industry in the United States. The USTA is the premiere forum for the small, mid-size and large companies of the local exchange carrier industry. The association represents more than 1,200 companies nationwide. Its members have total revenues from domestic telecommunications operations of over $100 billion.

The goal of the USTA is to promote the general welfare of the telephone industry, to collect and disseminate industry information and to provide a forum for the discussion and resolution of issues of mutual concern. USTA has been serving its member companies for more than a century.

Telecommunications Research & Action Center

Save $$
Click **HERE** To Find The Best Rates For Your Calls!

WebPricer

Order TRAC Publications

Recent News

● **Holiday Alert: Do A Long Distance Check-Up Now!** - December 16, 1998

● **Small Business Alert: Long Distance Costs Have Decreased** - Nov 5, 1998

● **Vacancy Announcement** - Nov 6, 1998

TRAC   www.trac.org

This site can help guide you through the maze of long distance calling plans available in the marketplace. TRAC, the Telecommunications Research & Action Center, is a non-profit organization dedicated to helping residential and business customers make educated long distance carrier choices. The centerpiece of this site is WebPricer, an automated tool that enables you to compare the price you will be charged for interstate calls by participating long distance carriers. You can also scroll through articles from TRAC's Tele-Tips newsletter to help you effectively manage your long distance expenses.

Telecommunications Resellers Association   www.tra.org

If you're looking for low cost long distance service, a reseller could be the answer. Many resellers buy excess capacity from the major long distance providers and sell it at a reduced price. The

Telecommunications Resellers Association (TRA) represents companies involved with the reselling of telecommunications services. This site provides useful information on switched and switchless resellers of domestic and international long distance services. You can also get information about the latest legal and regulatory issues, news updates and links to any TRA member's website.

Association for Local Telecommunications Services www.alts.org

The Association for Local Telecommunications Services (ALTS) is a national trade organization representing over 100 Competitive Local Exchange Carriers (CLECs). It provides a voice for CLECs looking to expand their business. At this site you can read up on ALTS filings, conferences and events. The site features a members-only area where CLECs can gain on-line access to the expertise of the ALTS staff and representation of competitive carrier's concerns in major federal proceedings. ALTS can also help in arranging visits for members to Capitol Hill, the Federal Communications Commission, or the Department of Justice.

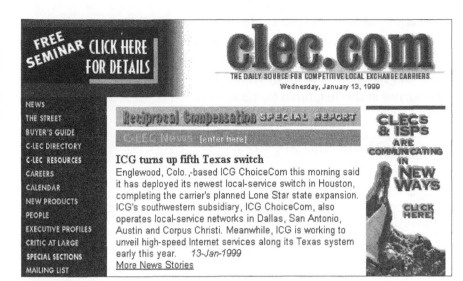

CLECinfo   www.clec.com

Are you looking for alternatives to your traditional local phone company? Everything about CLECs and their potential subscribers can be found on this site. Searching for CLECs is made easy with this search engine. You can look up carriers by city, state, or even chief executive's name. Included are top news stories on CLECs, and a page of links to equipment suppliers, billing vendors and more.

Telestrategies   www.telestrategies.com

Telestrategies is a leading sponsor of telecommunications industry conferences, seminars and trade shows. They also publish the monthly Billing World on-line newsletter (see their screen under On-Line

Publications above). Recent conferences include Telecommunications Taxation, OSS (Operational Support Systems), Federal Telecommunications Conference, and Marketing and Selling Telecom IT Solutions. You'll also find links and information about several other seminars and trade shows at this helpful, informative site.

A Listing Of Carriers' Websites

Company	Homepage
AT&T	www.att.com
Aliant Comm. Intl.	www.aliant.com
Alltel	www.alltel.com
Ameritech	www.ameritech.com
Associated Group	www.agrp.com
Bell Atlantic	www.bellatlantic.com
BellSouth	www.bellsouth.com
Century Tel.	www.centurytel.com
Cinn. Bell	www.cinbellinc.com
EXCEL	www.exceltel.com
Executive Telecard	www.etcard.com
Frontier	www.frontiercorp.com
Group Long Distance	www.gldnet.com
GTE	www.gte.com
IDT Corp.	www.idt.net
Inacom	www.inacom.com
IXC	www.ixc-comm-net
MCI	www.mci.com
Pacific Bell	www.pacbell.com
Qwest	www.qwest.com
RCN	www.rcn.com
SBC Comm.	www.sbc.com

Company	Homepage
Sprint	www.sprint.com
Teleglobe	www.teleglobe.com
Teligent	www.teligentinc.com
TelSave	www.telsave.com
TotalTel USA	www.totaltel.com
Trescom Intl.	www.trescom.com
USLD Comm.	www.usld.com
USWest	www.uswest.com
Williams Network	www.willtales.com
Winstar Communications	www.winstar.com
Worldcom	www.wcom.com

The FCC and State Regulators

Click here to display a **Text-Only** version of the FCC Home Page.

The Federal Communications Commission (FCC)   www.fcc.gov

The FCC's website offers an enormous amount of information about the latest regulatory and legislative issues pertaining to telecommunications. Interested in number portability, market share, or have a question about cellular towers? Check out this site. It has sections for each of the FCC's bureaus.

Alabama Public Utilities Commission	www.state.al.us
Alaska Public Utilities Commission	www.state.ak.us
Arizona Corporation Commission	www.cc.state.az.us
Arkansas Public Service Commission	www.state.ar.us
California Public Utilities Commission	www.cpuc.ca.gov
Colorado Public Utilities Commission	www.dora.state.co.us
Connecticut Dept. of Public Utility Control	www.state.ct.us
Delaware Public Service Commission	www.state.de.us
Florida Public Service Commission	www2.scri.net
Georgia Public Service Commission	www.psc.state.ga.us
Idaho Public Utilities Commission	www.puc.state.id.us
Illinois Commerce Commission	www.state.il.us
Indiana Utility Regulatory Commission	www.state.in.us
Iowa Utilities Board	www.state.ia.us
Kansas Corporation Commission	www.kcc.state.ks.us
Kentucky Public Service Commission	www.state.ky.us
Louisianna Public Service Commission	www.lpsc.org
Maine Public Utilities Commission	www.state.me.us
Maryland Public Service Commission	www.psc.state.md.us
Mass.Dept. of Telecommunications and Energy	www.magnet.state.ma.us
Michigan Public Service Commission	ermisweb.state.mi.us
Minnesota Public Utilities Commission	www.state.mn.us
Mississippi Public Service Commission	www.mslawyer.com
Missouri Public Service Commission	www.ecodev.state.mo.us
Montana Public Service Commission	www.psc.mt.gov
Nebraska Public Service Commission	www.nol.org
Nevada Public Utilities Commission	www.state.nv.us

New Hampshire Public Utilities Commission	www.puc.state.nh.us
New Jersey Board of Public Utilities	www.njin.net
New Mexico State Corporation Commission	www.state.nm.us
New York Public Service Commission	www.dps.state.ny.us
North Carolina Utilities Commission	www.ncuc.commerce.state.nc.us
North Dakota Public Service Commission	pc6.psc.state.nd.us
Ohio Public Utilities Commission	www.puc.ohio.gov
Oklahoma Corporation Commission	www.occ.state.ok.us
Oregon Public Utility Commission	www.state.or.us
Pennsylvania Public Utility Commission	puc.paon-line.com
Rhode Island Public Utilities Commission	www.ripuc.org
South Carolina Public Service Commission	www.psc.state.sc.us
South Dakota Public Utilities Commission	www.state.sd.us
Tennessee Regulatory Authority	www.state.tn.us
Texas Public Utility Commission	www.puc.texas.gov
Utah Public Service Commission	web.state.ut.us
Vermont Department of Public Service	www.cit.state.vt.us
Virginia State Corporation Commission	dit1.state.va.us
Wash.Utilities and Transportation Commission	www.wutc.wa.gov
West Virginia Public Service Commission	www.state.wv.us
Wisconsin Public Service Commission	badger.state.wi.us
Wyoming Public Service Commission	psc.state.wy.us

Chapter Summary

Like most search sessions on the Internet, looking for information to help audit your telecommunications billing can get complicated, time consuming and frustrating. The process improves with practice and familiarity with the sites that best meet your needs.

The Internet is an invaluable resource that can be used to keep your telecommunications environment both economized and in good working order. Whether you use it yourself or rely upon an auditing professional to use it for you, you can benefit enormously from the telecommunications information it has to offer.

9

Processing
and
Validating
Technology
Invoices

Organizations with active and complex telecommunications environments commonly wade through a sea of monthly invoicing trying to confirm work order completions, pricing accuracy and correct application of credits. Not knowing where to start often causes this labor-intensive effort to slip lower on the priority list. When left to languish too long, this task simply becomes impossible. At this point, the best you can hope for is to identify only current billing errors and obvious mistakes. Those overlooked errors and mistakes can go uncorrected for years.

Many organizations look for ways to streamline and automate their telecommunications invoice processing methodology without increasing resource requirements. Outsourcing this entire process is really the only way to accomplish this. An effective outsourced invoice processing service can improve your operations support service capability, and create a means by which you can analyze and validate telecommunications expenses faster and more accurately across multiple vendors. The outsource service provider should undertake the task of validating monthly invoice data and reconciling work orders and inventory against current billing. With your authorization, they should interface with your vendors to resolve all billing issues on a continuing basis. All invoice database administration and invoice approvals should be coordinated and approved by you. Once invoices have been approved for payment, the appropriate information should be forwarded to your Accounts Payable system via hard-copy check requisition forms or electronic advice for remittance. Care should be taken to ensure these charges are correctly allocated across your organizational hierarchy for posting to the general ledger.

The Difficulty with Processing Invoices

Telecommunications invoice processing, if done manually, can require hundreds of hours a month for many organizations. In today's deregulated, multi-vendor telecommunications environments, this process is intensifying. It is not uncommon for organizations to use several local telephone companies and long distance carriers. It is also typical for them to use many different independent suppliers of telecommunications products and services. This can result in a staggering number of invoices each month and an ever-increasing risk that you will be overpaying for telecommunications services.

Most organizations start by trying to keep their telecommunications costs low, but as they grow, so do their telecommunications requirements. It is easy to lose track of what systems are in place and what services exist. Organizations tend to lose control of what they are paying for, who is using the services they have, what cost efficient alternatives are available and what budgetary guidelines should be used for controlling these expenses.

New technologies, particularly Internet and Web-based services, are adding to the complexity. Many of these new technologies, though convenient, require the installation of new transmission lines, hubs, routers and various other components related to their architecture. Although these technologies save time, they also tend to increase the telecommunication invoice count, with both recurring and non-recurring charges.

An In-house Solution

Telecommunication expense management is a very complicated and at times bewildering process. One way to address complex billing issues is to have a dedicated in-house staff that works only on telecommunications invoice processing, reconciliation and related expense management. A unit like this would need to include staff which are knowledgeable about USOCs, FIDs, and the interpretation of rates, tariffs and special billing contracts. This is a very expensive solution and not a very efficient way to manage this operation support service. Units of dedicated employees like this tend to become subject matter experts in narrow areas of expertise. As people transfer or leave, their expertise is lost and a gap appears in the overall process. No training program can adequately recreate the lost expertise in a timely fashion.

An Outsourced Solution

A true Invoice Processing and Validation Service provider should provide a proactive approach to controlling information technology billing expenses and be a resource to manage related information. These service providers can effectively manage complex telecommunications billing structures from multiple vendors. A service provider should be able to eliminate unnecessary paper handling and manual data entry, creating savings for your organization while reducing the propensity for human data entry error. Process enhancements will allow your existing staff to focus on creating cost avoidance measures or bulk buying opportunities, which can be realized as hard dollar savings. Automation of existing manual invoice processing and the implementation of invoice screening parameters can help focus an organization's staff on strategic issues, also resulting in hard dollar savings. They allow you to take advantage of EDI (Electronic Data Interexchange), thereby eliminating your involvement with data collection, number crunching, information distribution and an IT infrastructure to support the process. Electronic billing can provide you with valuable management, exception and financial reporting and information delivery. This information should be available from a database, or via the Internet. Such information can help negotiate better rates/contracts, reconcile vendor payments, monitor trending and facilitate budget analysis.

What Is Involved

You will need to have an initial meeting with your Invoice Processing Service provider and designate appropriate project manager(s) and other designees as this service crosses internal department boundaries. It is important to establish and confirm overall invoice processing and validation project plans, define objectives and establish priorities. Your service provider must have a clear understanding of your organization's invoice receiving, approval, allocation and payment processes to be established and followed during the performance of services.

Expect to provide your service provider a complete listing of your authorized vendors, respective billing account numbers along with a list of your vendor's contacts and special contract information specific to your custom billing arrangements. The invoice processing service provider will need a Letter of Agency to act on your behalf with your vendors to obtain pertinent information and fulfill their service obligations to your organization. Be sure that you or your service provider document and define the Invoice Processing Service methodology developed for you, along with policies and procedures for invoice remittance. Expect this document to be continually modified in dynamic environments.

An outsourced provider of Invoice Processing and Validation services should perform all the work necessary to establish working databases and redirect your paper invoices to their processing facility. Upon receipt, an invoice should be inventoried into their invoice management database and put through a combination of pre-determined field comparisons; an invoice must meet certain validation criteria in order to be processed. Having pre-established criteria ensures the avoidance of payment covering, duplicate invoices, invoices exceeding a fixed dollar or percentage threshold, bills with invalid "Remit To" addresses, bills reflecting a new vendor account number or invoices not belonging to your organization. The service provider should analyze the various vendors' paper invoices and number of items to determine if bill consolidation makes sense. If it does, they should interact with your vendors to consolidate appropriate billing. Once established, the invoice information should be captured in available electronic platforms. EDI would be the most preferable as it has information rich content and can be used for management and exception reporting. EDI billing allows your service provider to more accurately validate and reconcile your monthly invoices.

Ensuring Accuracy

Invoice validation efforts should be performed concurrently with the processing of invoices for payment. Custom analysis/validation specific to your organization's environment should be performed. Consistent with a methodology for conducting invoice validations, your invoice processing services should include (a) historical reviews, (b) on-going compliance to ensure that your service provider correctly remedies past overcharges, and (c) financial management controls to monitor invoices going forward.

- *Custom Analysis/Audit Application*

   In conjunction with your service provider, outline specific billing items and variances that will require monthly validation. Keep in mind the custom analysis/audit effort should be designed to extract billed items that despite being billed correctly according to applicable tariffs, violate your organization's policies and procedures. Such items can include fraudulent and abusive charges, late payment fees, 900/976 type calls, third party billing, USF charges and incorrectly selected long distance carrier calls. Your service provider must be able to identify these exceptions and interface with your appropriate vendor(s) to secure financial adjustments as required. It is important that you work with your service provider to develop and implement permanent cost avoidance measures to correct these issues going forward.

- *Work Order Reconciliation*

   Invoice reconciliation is a key element to ensuring a correct invoice. Be sure you can capture work order activity and associated inventory information to facilitate such invoice reconciliation efforts. This positions you to ensure that all Move/Add/Change activity has been properly executed and billed appropriately. In the absence of such information, electronic billing would enable you to prepare monthly-billed line item comparisons for review. If any anomalies are detected as part of the work order to invoice reconciliation, your service provider should now be best positioned to obtain proper adjustments for incorrectly charged work order activity, or invoices that cannot be reconciled.

- *Historical Bill Audits*

   A comprehensive service provider must be able to ensure invoice charges are not only correct today, but that historical billing has been applied correctly. They will need to review your telecommunications charges to determine if incorrect billing and/or overcharges exist. They should be positioned to audit your invoices with respect to the application of correct rates, tariffs, special contracts, taxes and surcharges. Once items are identified as incorrect, your service provider should document the findings and interface with your vendors to resolve billing anomalies and obtain applicable revenue recoveries as refunds or credits on your behalf.

Allocation and Chargeback

Once an invoice is validated and approved for payment, it then must be properly allocated in accordance with your organization's chargeback methodology. This is often performed in accordance with your organization's requirements and policies for maintaining financial controls. Your allocation methodology should be established in your service provider's invoice management database. It should be managed each month and have the capability to be modified to ensure that all invoices promoted for remittance comply with your organization's unique allocation and chargeback methodology.

Getting the Invoice Paid

Now that the invoice has been validated and allocated, you can now have the comfort level that the invoice can be properly remitted for payment. You will need to work with your service provider to develop and implement a payment transmittal process to your accounts payable department. This will require the establishment of a single payment transmittal design to be achieved by hardcopy or electronic transfer that complies with your organization's financial policies and procedures for invoice remittance, as well as, general ledger posting. There are significant cost savings if an electronic payment process can be achieved. This allows your organization to refocus accounting staff from data entry of approved paper invoices, to more strategic and productive initiatives. Regardless, all pertinent information for invoice remittance and allocations should be contained in the designed transmittal process.

Once you and your service provider establish coordination with your accounts payable department, they should ensure invoice payments are made within proper remittance schedules to vendors. An added check and balance feature to consider is having your service provider forward prepayment approval files for your review prior to sending the information to your accounts payable department. A sophisticated service provider will have the capability to receive your payment advice and populate their invoice management database with such details. This eventually allows them to work with your vendors to ensure that debits and credits are properly applied.

The Benefits

Armed with a powerful resource like an automated Invoice Processing and Validation service, you are able to:

- Allocate costs to proper cost centers
- Take advantage of EDI
- Maintain control of asset inventories
- Perform general ledger and accounting related analyses
- Monitor vendor contract compliance

- Resolve invoice management problems
- Minimize your involvement in report production/data distribution
- Ensure billing accuracy
- Reduce costs
- Automate posting of IT expenses directly to the General Ledger
- Focus staff on more strategic issues.

When correctly implemented and deployed, Invoice Processing and Validation services reduce and control expenses through better awareness of costs by identifying to users how those costs impact an organization's overall information technology and telecommunications expenditures.

There are a limited number of service providers offering invoice processing and validation type services. A comprehensive organization should take on the task of managing your entire telecommunications invoicing environment, including associated costs. This should include arranging to have all invoices redirected to their facility and then ensure the invoice data is populated within a database for processing routines. These routines include audits, validations, reconciliation, allocations and exception/management reporting.

Access To Information

As the client, you should be provided access to your information database. You can then call forth any invoice and examine any record for any reason. Invoice Processing services should include the delivery of invoice, call detail and other types of information via the Internet or a company Intranet. These platforms reduce the excessive cost of paper reports, thereby giving you direct, paperless access to the specific information needed. Internet delivery of data will help you manage invoice expenses faster, more accurately and across multiple vendors in ways that are currently difficult, if not impossible to achieve by other available methods providing such information. You need this for financial analysis, cross-references, vendor comparisons, and exception management identification. Overall, the information must enhance your ability to effectively manage technology and telecommunications operations functions.

A cost effective strategy for managing information technology expenditures will allow you to control and monitor information technology billing expenses and related resources without incurring human resource expenses. The good invoice processing service provider takes on the brunt of the grunt work. You should have access to all information through concise and comprehensive reporting vehicles.

The Key to Success

Collaboration is the key to making this service work. The companies that process technology invoices have to be able to work with your existing staff to blend their expertise with the knowledge resident within your company. Working together as a whole unit maximizes the potential to process invoices efficiently, as well as, implement billing validation controls effectively. A good service provider focuses directly on this goal. Rather than attempting to break up the existing system and start from scratch, it should concentrate on ways to create a stronger organization starting with your existing bill paying infrastructure. It should eliminate the burden of maintaining a full-time staff dedicated to invoice administration and allow you and your staff to pursue other, more pressing telecommunications issues.

Chapter Summary

Telecommunications expenses in large organizations are experiencing multifaceted growth. Internal expansion and the diversity of regulated and unregulated telecommunications products and services being offered today affect them. Keeping track of these expenses involves either the creation of an internal organization or unit dedicated to invoice reconciliation or outsourcing this task to a qualified vendor. Left unchecked, these expenses can accumulate to a point where they hide perpetuating errors and mistakes. They need constant monitoring and control. Industry statistics, which point to the fact that 80% of telecommunications billing is wrong, lend credence to this belief. Outsourced services can save time and be a cost-effective way to improve cumbersome operation support services while reigning in out-of-control expenses and controlling the growing demand for telecommunications and information technology services.

10

Call Accounting, Phone System Security, Toll Fraud and Toll Abuse

Companies must take proactive steps to protect their network from unauthorized use. The first and most basic step you can take is to implement and use a Call Accounting System. A Call Accounting System is a computer-based management system designed to capture, package and report call record detail (CDR). It stores information regarding call characteristics, i.e., carrier, length of call, time of day and the number called. It is the vehicle through which call pricing is verified. With some systems, information concerning incoming calls is also recorded.

There are several management benefits associated with call accounting systems, some of which are described below.

1. *Controlling Telephone Use and Abuse* – Telephone use can be abused by employees. Telecom managers can use management reports from their call accounting systems to pinpoint areas where excessive non-business calls are originating and to also isolate areas where other abuses are occurring.

2. *Cost Allocation* - Reports from call accounting systems can be used to drive expense reports directly to those departments responsible for them.

3. *Client Billing* - If clients are billed for time spent with them on the telephone, or if professionals bill you for telephone time, you can double check the accuracy of this billing against call accounting reports.

4. *Personnel Development* - Call accounting reports can be used to monitor and motivate telephone sales and telemarketing employees. They can also be used in productivity reviews when it is suspected employees are spending too much time on personal calls.

5. *Telecommunications Management* - Data for call accounting systems can support evaluations of overall telecom environments, assess least cost routing logic, and to ensure overall billing accuracy.

Call accounting systems are typically PC-based or stand-alone units. They read and assemble stored data at preset intervals, either directly from disk, across an attached LAN, or by polling the data recorder/buffer box. Many systems use buffer boxes, which can be set up several ways. The buffer box can be on-site sitting next to a switch, or, if you have several sites, can be at each remote site. Buffer boxes can be set to send call records to your call accounting PC (via modem) at certain times, i.e., when it is full, when you initiate an uploading command, etc.

Once your call accounting software has captured all appropriate records, it turns them into standard or customized reports. The software consults tables (V&H coordinates, tariff, etc.), and applies algorithms and other factors to calculate actual call cost. Then, depending upon your standard and customized report requirements, it releases reports by Extension, Department, Longest/Most Expensive Calls, Frequently Called Numbers, Trunk Report, and Toll Fraud Detection. These reports will typically list the extension that made the call, the trunk it went out on, how long the call lasted, how much it cost, the number dialed and how much each department spent. Customized reporting, available on many call accounting systems, lets you design your own reports to obtain the detailed breakdown of your communications in the format that is best for you.

Reports can be saved as a file on a PC, e-mailed, and/or sent to printers. Some LAN-based call accounting systems let authorized people view reports on PCs attached to the LAN/WAN. And increasingly, more systems are letting you access reports via the Internet/Intranets with a browser. High-end call accounting systems can monitor multiple switches. Some work in conjunction with other software modules to monitor other network infrastructure components like Virtual Private Networks (VPNs) and Internet connections. By utilizing Web browsers and the Internet, call accounting systems provide a familiar interface and convenient way to access reports. Travelling managers can get information they need on their laptop from their hotel room. Some systems can be set up to generate reports in HTML format. Then, at predetermined times, the reports are posted to your Web site. You log on, enter a pass code, and the call records are available to you.

Phone System Security and Preventing Toll Fraud and Toll Abuse

Slamming is the practice in which a customer's local toll or long distance service provider is switched without their permission. It is the number one complaint of telephone customers to the FCC.

You can prevent slamming by placing a freeze on your account. Simply call up your local telephone company and request a freeze on all PIC changes. The freeze must be placed on all of your telephone lines. A freeze will only allow an authorized representative of your company to change the PIC. You can verify that the freeze has been implemented by requesting a CSR after the freeze is called in. After each telephone number on the CSR you will see the following: /PIC MCI/PCA FN, 7-19-99. The FN stands for freeze.

You can call 1-700-555-4141 to verify the long distance carrier on any line. The verification call must be made from the actual line you want to verify. Call your local telephone company if you have been slammed so you can be switched back to the proper long distance carrier. Report the slamming to the FCC.

Beyond the obvious day-to-day management benefits inherent in call accounting systems, there is a darker, seamier side to all this that demands equal vigilance. More than ever before, hackers and others who seek to defraud telephone systems are invading corporate telecommunications systems. Toll fraud is more commonplace today than ever before. Remote PBX toll fraud is the unauthorized use of networks by hackers who are physically outside your premises. It is estimated that businesses lose millions of dollars each year to PBX toll fraud.

Many call accounting systems, including those utilizing buffer boxes, can be used to monitor and prevent toll fraud and toll abuse. Call accounting systems can be set up with various thresholds so that if a call goes over a preset dollar amount a supervisor can be alerted. Costly long distance calls that did not originate from within your office, but went through your PBX, can mean someone hacked your switch and illegally obtained dial tone. Buffer boxes can be programmed to read incoming toll-free call information and send a message (typically via pager) if an undesired area code is dialing into a switch connected to the buffer box.

Signs of a Hacker

The following are some indications that could mean your phone system is being hacked.

- Authorized employees calling in for messages at night receiving busy signals because all ports are in use. Hackers are likely to break-in late at night via the voicemail system and lock everyone else out.

- There are too many busy lines for the amount of traffic your business should generate at a particular time of day or night.

- CDR reports show lots of calls during abnormal times, such as over holidays, weekends, etc.

- Switch reports show multiple, failed log-in attempts to a system.

- Unexplained surges in system use.

- Long holding times.

- Reports showing calls to unrecognizable area codes and/or country codes.

- Your voicemail codes have suddenly, and unexplainably, changed.

- Incoming calls being transferred to a particular extension that has been set up for call forwarding.

Ways to Reduce Your Exposure to Toll Fraud

Hackers can break into your PBX via DISA (Direct Inward System Access), remote maintenance ports and through your voicemail system. The FCC has ruled that the customer, not the long distance carrier, is responsible for payment of the fraudulent calls (Chartways vs. AT&T). That is largely due to the fact that these types of calls appear on your bill as directly dialed from your PBX. You are only responsible for the first $50.00 of usage charges on fraudulent calls made via stolen calling cards.

Hackers use other means of stealing dial tone as well. One method is the "dumpster dive". They look through trash, searching for records that might contain pass codes and PINs. Make sure documents with such security information aren't left lying around or simply thrown out. Shred them. Hackers also try to talk their way into your switch. They use various ruses, sometimes pretending to be service technicians. They'll call up and ask for certain pass codes or for access to a line, under the guise of being an employee of the local Telephone Company. Instruct employees not to give out information, or dial tone, to anyone. If you aren't sure about someone claiming to be a technician, you should get their name and number, hang up and call them back to check.

Recently, prisoners were found to be calling people and claiming they were service technicians conducting tests on telephone lines. They told the person they called that to complete the test, he or she should press '9', '0', the pound sign, '#', then hang up. By pressing this sequence of buttons, you are providing the caller with full access to your telephone line. They can place long distance calls over your line, at your expense. Again, ask the caller in situations like this for their number, hang up and call back.

Telecommunications managers can take several steps to reduce toll fraud and toll abuse. Start by talking with your switch and voicemail vendors about security issues. Find out what types of built-in security features are on your systems. Several major long distance carriers also offer security software.

Review and document all security measures taken by your company. Most importantly, monitor these measures. Conduct an audit to ensure the following steps have been taken to protect your company's voice network.

- Perform an audit of all your telecommunications facilities. Remember, you cannot protect telephone lines that are not documented. Disconnect all unused POTS and special service lines (i.e. PBX extensions).

- Prepare a network diagram of all your facilities. Your network security is only as good as the weakest node connected to it. If, for example, you have tie lines connecting PBXs at different locations, you must secure all of the locations. If one PBX is compromised then all your locations are at risk.

- Assign a responsible department to all telephone facilities. Each department should be given a list of facilities that it is responsible for. Each department should also be required to review its telephone facility list to ensure that all facilities are needed. Each department then becomes responsible to notify the telecommunications department when they downsize and no longer need certain facilities.

- Tracking - You must require all departments to budget and estimate monthly telephone costs. Actual totals should be compared to estimates. Reports should be generated that track trends by department. Telecommunications is an expense that should be borne by the user of the services.

- Blocking - You can block 900, 976 and 700 – type calls, etc., through your local telephone company. Most will charge you a one-time fee to block these calls (usually about $25 to $30). You can also block third number calls and collect calls on each individual telephone line number. A line can also have call screening placed on it. Hospitals and hotels typically utilize call screening. It prevents you from making a directly dialed call from a particular line. An operator will intercept all calls and force you to place the call collect or bill it to a credit card.

Blocking of 900, 976, 700 and 500 series calls will appear on the CSR. A Bell Atlantic CSR will list blocking as follows (The USOC for blocking is RTVXC):

RTVXC   /LCC ABB (Blocking Service Charge).

Blocking of third number and collect calls will appear on the CSR as follows:

74.TKNA,718,555,1234/TBE A/PIC ATT

/TBE A blocks both collect and third number calls on a particular line.

/TBE B blocks only third number calls.

/TBE C blocks only collect calls.

- Blocking by the long distance Telephone Company. You can have your long distance carrier assign access codes to particular line numbers. They can also be assigned by department or to each employee. Certain access codes can be restricted from dialing certain area codes. This type of blocking only works for calls placed over your primary carrier's network. If employees or visitors dial the access code for other than your primary carrier, they will circumvent this blocking.

- Blocking at the PBX. Your PBX can block access to certain area codes by telephone line or extension by class of service assignments. Certain lines can be restricted while others are left unrestricted. You can also restrict by time of day. For example you can block all calls made from a telephone line after 6 P.M. Calls to costly long distance area codes and country codes should be blocked whenever possible.

- All remote maintenance ports and inbound modem lines should be disconnected when not in use or they should be equipped with call back devices. Also called Callback Modems, these answer when you remotely dial into a system and request a password. After you enter the password you hang up. It then calls you back at a predetermined, authorized phone number.

- Disconnect or restrict DISA access. The most common way hackers compromise your voice network is by breaking your DISA passwords.

- Institute system access password security measures. Make sure your initial PBX default password is frequently changed. Change PBX access passwords quarterly. Don't use obvious or easy to guess passwords based upon names, addresses, etc. With any numeric codes, don't use patterns that make it easy to remember, like repeating numbers, geometric patterns, etc. Codes and combinations can be generated by automated hacking software. Delete access codes for employees that leave the company ASAP.

- Prepare a Toll Fraud Prevention Plan. Document and review all prevention measures taken to protect your voice network. Update this plan as the business changes.

- Prepare a Toll Fraud Contingency Plan. This plan documents the steps you will take if PBX toll fraud is detected. It will detail the steps you will take if a hacker has successfully broken into your voice network. This plan should include the 24-hour numbers of your long distance carrier, maintenance vendor and the home phone numbers of at least three responsible employees. In addition to your primary carrier's 24-hour number have the 24-hour network security number for AT&T, MCI and Sprint handy in case the hacker dials an access number to bypass your primary carrier. These companies allow casual calling, and hackers typically place fraudulent calls over a variety of long distance carriers.

Protecting Auto Attendants and Voicemail

Hackers can also break into your auto attendant/voicemail to steal dial tone. Since voicemail answers outside calls and provides menus and avenues for callers to navigate your system you need to make sure it's secure. Hackers have been known to break into a voicemail system and leave messages in unused mailboxes for other hackers. Their messages include codes, passwords and access methods for breaking into your voicemail system. If you manage an auto attendant/voicemail system, keep this security checklist in mind.

- Block transfers to numbers not defined as extension numbers or auto attendant choices. Don't let callers get an outside line by pressing "9" or some other trunk access number. Set up your voicemail lines to only answer calls, not dial out.

- Prevent unauthorized use of server storage space. Don't let callers leave messages in uninitialized mailboxes (boxes that have been created but aren't being used). This prevents unauthorized callers from using mailboxes and hearing sensitive company information sent out over voicemail by logging onto an unused mailbox. It also prevents hackers from doing cute things like leaving messages for other hackers, messages that could contain access codes to your switch, personal information about employees heard on their mailboxes, etc.

Telecommunications Expense Management

Appendix A

Commonly Encountered USOCs

USOC	Explanation	LEC which utilizes the USOC			
1AQ	ISDN-message serv	Bell Atlantic			
1D91X	FX local channel-primary channel		BellSouth		
1FB	Business line-flat rate		BellSouth		
1FL	Business line-flat rate			Southwestern Bell	
1FR	Residential line-flat rate		BellSouth		
1L9FX	Foreign exchange mileage		BellSouth		
1LDP1	Mileage charge	Bell Atlantic			
1LPDA	Mileage charge	Bell Atlantic	BellSouth		
1LDPZ	Mileage charge	Bell Atlantic	BellSouth		
1LNO1	Interoffice channel mileage, fixed rate, 0-8 miles		BellSouth		
1LNOA	Interoffice channel mileage, each airline mile, 0-8 miles		BellSouth		
1LS1H	Suburban serv area boundary-one-pty line			Southwestern Bell	
1LVFR	Mileage-each 1/4	Bell Atlantic			
1LZHA	Mileage-pvt line channel		BellSouth		
1MB	Business line-measured rate	Bell Atlantic			Pacific Bell
1PQWA	Megalink/lightgate, analog trunk		BellSouth		
1RSEX	DDSII/enterprise/frame relay	Bell Atlantic			
1S8	Business line, economy service option		BellSouth		

USOC	Explanation	LEC which utilizes the USOC			
1ZJ	Business line, standard service option		BellSouth		
1YEFS	Interoffice digital channel	Bell Atlantic			
3LBAS	Intralata private line service, interoffice channel		BellSouth		
3LBBS	Intralata private line service, interoffice channel		BellSouth		
3LN1S	Mileage charge	Bell Atlantic			
3LN2Y	Mileage charge	Bell Atlantic			
3LNGJ	Mileage charge	Bell Atlantic			
5OZ	Channel protector	Bell Atlantic			
742SA	Entrance bridge	Bell Atlantic			
7TTEX	Touch-tone line	Bell Atlantic			
888WS	WATS-cross reference-inward or outward line			Southwestern Bell	
9LA	FCC charge for network access for additional line		BellSouth		
9LM	FCC charge for network access		BellSouth		
9PZB1	FCC line port charge	Bell Atlantic			
9PZDD	FCC line port charge	Bell Atlantic			
9PZSD	FCC line port charge	Bell Atlantic			
9PZ	FCC line charge	Bell Atlantic			
9ZEU4	Access interstate calling multi-line business				Pacific Bell
9ZP	Subscriber intrastate line charge, business		BellSouth		Pacific Bell
9ZR	FCC line charge	Bell Atlantic	BellSouth	Southwestern Bell	
9ZRB1	FCC line charge	Bell Atlantic			
9ZR22	End user access charge (EUCL)-business			Southwestern Bell	
A6T	Modem	Bell Atlantic			

USOC	Explanation	LEC which utilizes the USOC			
ACB	Area calling service-business-economy		BellSouth		
AH8	Telecommunications relay service		BellSouth		
ALN	Additional line	Bell Atlantic			
ALS	Additional message	Bell Atlantic			
ANPCD	Billmanager-monthly	Bell Atlantic			
AP2	Multiple bill charge	Bell Atlantic			
ASB	Area calling service-business standard		BellSouth		
AS3SL	Software line	Bell Atlantic			
AS3SS	Software line	Bell Atlantic			
AWS	Customer alerting enablement on POTS			Southwestern Bell	
AXGHX	Plexar express-access line intercom loop-0-2 mile			Southwestern Bell	
BBC	Block busyconnect announcement		BellSouth		
BCF	Busy verification	Bell Atlantic			
BCR	Call return blocking		BellSouth		
BM2	PBX meas addl trunk line 10.90 per month				Pacific Bell
BKDXA	Directory assistance	Bell Atlantic			
BQ9	Intralata private line service, voice grade bridge		BellSouth		
BRD	Repeat block dialing		BellSouth		
BSX	Bell Atlantic calling card	Bell Atlantic			
BSXUP	BellSouth calling card		BellSouth		
BW1	SRV conn labor chrg for install of addl jacks and/or wire runs				Pacific Bell
BWC	SRV conn labor chrg for install of first jacks and/or wire runs				Pacific Bell

USOC	Explanation	LEC which utilizes the USOC			
CCOEF	Clear channel capability extended superframe format		BellSouth		
CKY5A	Comkey 2152 answering position station line charge		BellSouth		
CLTCX	Additional listing	Bell Atlantic			
CLTLX	Additional listing	Bell Atlantic			
CLT	Addl white page lstg(s): bus 3.50 per month				Pacific Bell
CON2X	Off premise extension	Bell Atlantic			
CON4X	Loop charge 4-wire	Bell Atlantic			
COPXX	PBX system, customer owned and maintained, flat rate		BellSouth		
CPERE	Customer owned connection equip, chargeable if telco provided		BellSouth		
CPERN	Customer owned connection equipment		BellSouth		
CPL2X	Off premise extension	Bell Atlantic			
CRDO3	Non touch-tone line credit		BellSouth		
CREX1	Custom toll restriction		BellSouth		
CREX4	Custom toll restriction		BellSouth		
CREX6	Custom toll restriction		BellSouth		
CREXK	Outgoing calls preceded by a 900 prefix			Southwestern Bell	
CROBA	Chargeback-business			Southwestern Bell	
CTG	Circuit termination charge	Bell Atlantic	BellSouth		
CYS	Common equipment	Bell Atlantic			
*CY	Sales tax				Pacific Bell
D1F2X	Off premise extension	Bell Atlantic			
D42CT	Switchway service	Bell Atlantic			
DO8	Intellipath - digital	Bell Atlantic			

USOC	Explanation	LEC which utilizes the USOC			
DOX	ISDN basic exchange secondary	Bell Atlantic			
DPN3L	Dataphone digital service		BellSouth		
DPN3S	Dataphone digital service		BellSouth		
DRS	Ringmaster 1-one ringmaster number with distrinctive ring		BellSouth		
DYDRS	Digipath ii digital	Bell Atlantic			
E3PDF	Call-pickup	Bell Atlantic			
E5E	Call forwarding ii	Bell Atlantic			
E6GDF	Call forwarding	Bell Atlantic			
E9GDF	Call forwarding	Bell Atlantic			
E9GPA	Call forwarding don't answer		BellSouth		
E2H	Consultation hold & calling	Bell Atlantic			
EABDF	Automatic call back	Bell Atlantic			
ELY2N	User transfer and conferencing		BellSouth		
EMW	Service connection message waiting indicator				Pacific Bell
EPH	Call pickup/hold	Bell Atlantic			
ER3	Call forwarding and 8 code speed calling		BellSouth		
ES7	Custom calling package	Bell Atlantic			
ES92A	Plexar express-flat-class of service			Southwestern Bell	
ESA	Call forwarding, call waiting, and 8 code speed calling		BellSouth		
ESC	Three-way calling	Bell Atlantic		Southwestern Bell	Pacific Bell
ESL	8 code speed calling		BellSouth		
ESM	Call forwarding	Bell Atlantic	BellSouth	Southwestern Bell	

USOC	Explanation	LEC which utilizes the USOC			
ESMDF	Call forwarding	Bell Atlantic			
EST	Number capacity	Bell Atlantic			
ESX	Call waiting	Bell Atlantic	BellSouth		
EVBPA	Busy call forwarding 4.20 per month				Pacific Bell
EVBPA	Call forwarding busy line		BellSouth		
EVD	Delayed call forwarding 16.80 per month				Pacific Bell
EVD	Predesignated telephone number per line equipped			Southwestern Bell	
EVD	Call forward number			Southwestern Bell	
EXF	Off premise extension	Bell Atlantic			
EZC2B	Plexar-ii class of service (restructured)			Southwestern Bell	
EZC2B	30 workable sta capacity			Southwestern Bell	
FALLX	Foreign directory listing		BellSouth		
FALLX	Directory listing	Bell Atlantic			
FDA	BellSouth diskette analyzer bill service		BellSouth		
FLK	Foreign listing-Kansas			Southwestern Bell	
FRN77	Credit for EUCL based on access lines			Southwestern Bell	
FVJ	Outside system-per plexar-ii station equipped			Southwestern Bell	
FVJ	Call forward number			Southwestern Bell	
FPG3X	ISDN basic exchange digital	Bell Atlantic			
FX5CX	Foreign exchange access, combination trunk, measured		BellSouth		
FXGTS	Foreign exchange service measured business line		BellSouth		

USOC	Explanation	LEC which utilizes the USOC			
FXSTS	Foreign exchange service		BellSouth	Southwestern Bell	
FZA	Plexar-ii-basic station (restructured)			Southwestern Bell	
GCE	Call forwarding busy line		BellSouth		
GCJ	Call forwarding don't answer		BellSouth		
HBQ	Deny repeat call	Bell Atlantic			
HBS	Deny repeat call	Bell Atlantic			
HCABL	High capacity circuit	Bell Atlantic			
HCABS	High capacity circuit	Bell Atlantic			
HSHCH	Hunting-circle-local exchange serv feature			Southwestern Bell	
HSHPT	Regular hunting-per line/terminal in the group			Southwestern Bell	
HSHPT	Carrier is sprint			Southwestern Bell	
HTG	Hunting/rollover service	Bell Atlantic	BellSouth	Southwestern Bell	
HTGLO	Grouping for local optional service		BellSouth		
HWJ	Feature-per station (restructured)			Southwestern Bell	
JHSV4	Volume control handset	Bell Atlantic			
JJM1X	25-line connector jack	Bell Atlantic			
JJS5S	Single-line jack	Bell Atlantic			
JJS8S	Series jack	Bell Atlantic			
JUJ	Indoor two-line miniature jack	Bell Atlantic			
JZ22J	Mileage per 1/4 mile	Bell Atlantic			
LLT	Additional directory listing-cross reference		BellSouth		
LPRFX	Channels activated b channel flat rate circuit switched voice/data		BellSouth		

USOC	Explanation	LEC which utilizes the USOC			
LSN1X	2b channel and id	Bell Atlantic			
LTBLB	Individ line isdn business-low vol access/digital subscrib line (DSL)		BellSouth		
LTG1X	CKT switched voice & data b channel	Bell Atlantic			
LTH1X	ISDN basic exchange alternate	Bell Atlantic			
LTH4X	ISDN basic exchange low speed	Bell Atlantic			
LTH5A	ISDN circuit switched voice	Bell Atlantic			
LTH5X	ISDN basic exchange circuit	Bell Atlantic			
LTH6X	ISDN basic exchange circuit	Bell Atlantic			
LTQ8Y	ISDN user profile flat rate, includes caller id		BellSouth		
LTRUB	ISDN indiviual line, 5ess, flat rate business		BellSouth		
LYE	PBX line identification exception		BellSouth		
M1GNC	ISDN interoffice channel per DSL		BellSouth		
M1GNM	ISDN interoffice channel per mile		BellSouth		
MBJ1X	Standard mailbox	Bell Atlantic			
MB21X	Mailbox mltple	Bell Atlantic			
MFD2X	Multiple feature credit or two features		BellSouth		
MJB	Special business account	Bell Atlantic			
MVPSL	Prestige service		BellSouth		
MVSPX	Intellidial-service	Bell Atlantic			
N2Q	ISDN 2-wire loop	Bell Atlantic			
NCO	Additional call offering	Bell Atlantic			
ND1	Non-listed did automatic intercept service, per number referred				

USOC	Explanation	LEC which utilizes the USOC			
ND4	PBX service, additional group of 20 DID numbers		BellSouth		
ND6	Sliding scale charge	Bell Atlantic			
ND8	STA numbers			Southwestern Bell	
NDA	Located switching systems-each additional 10 stations			Southwestern Bell	
NDT	PBX service, BellSouth cmrs local loop did trunk termination	Bell Atlantic	BellSouth	Southwestern Bell	
NDV	PBX service, DID group of reserved station numbers		BellSouth		
NDW	Sliding scale charge	Bell Atlantic			
NDZ	Sliding scale charge	Bell Atlantic	BellSouth	Southwestern Bell	
NFB	Traffic record	Bell Atlantic			
NGD	24-port number group	Bell Atlantic			
NLE	Listing-not printed in directory, no charge		BellSouth		
NLT	Listing-not printed in directory		BellSouth		
NNS1X	Area calling srv, nar, megalink, per line or trunk, inward only		BellSouth		
NNSOX	Area calling srv, nar, megalink, per line or trunk outward only		BellSouth		
NP3	Listing-not in directory or directory assistance		BellSouth		
NPU	Non-published service	Bell Atlantic	BellSouth	Southwestern Bell	Pacific Bell
NQM	Megalink channel srv nar, per flat rated line or trunk both ways		BellSouth		
NQP	Megalink channel srv nar, per flat rated line or trunk, incoming		BellSouth		

USOC	Explanation	LEC which utilizes the USOC			
NRASQ	Circuit termination charge	Bell Atlantic			
NSP	Return, call blcoker, auto redial, priority call			Southwestern Bell	
NSR	Carrier is AT&T			Southwestern Bell	
NST	Call trace			Southwestern Bell	
NUM	Standard feature-per station-restructured			Southwestern Bell	
NV8	Class feature-auto redial-usage sensitive			Southwestern Bell	
NV9	Class feature-call return-usage sensitive			Southwestern Bell	
NV9	Carrier is AT&T			Southwestern Bell	
NW1	Network interface	Bell Atlantic		Southwestern Bell	
NW1O1	Network interfacing -outside single line		BellSouth		
NWL	Private line number of working loops		BellSouth		
NYZAA	Contract package	Bell Atlantic			
NZP	ISDN basic exchange display	Bell Atlantic			
OBM	Bill plus service establishment			Southwestern Bell	
ODKPS	Feature-per system (restructured)			Southwestern Bell	
OLKBX	ATOD-res/bus-intrastate			Southwestern Bell	
OSW25	Discounted calling plan, watssaver service, 25 hour plan		BellSouth		
OUV	Local usage discount plan	Bell Atlantic			
P1JAX	Intralata private line service, sub-voice grade local channel		BellSouth		

USOC	Explanation	LEC which utilizes the USOC			
P1JHX	Intralata private line service, sub-voice grade local channel		BellSouth		
P2JHX	Intralata private line service, voice grade local channel		BellSouth		
P2JMX	Intralata private line service, voice grade local channel		BellSouth		
P2JQX	Intralata private line service, voice grade local channel		BellSouth		
PENM1	Service connection ISDN installation appointment				Pacific Bell
PENSL	Service connection ISDN installation delay				Pacific Bell
PFSAL	Private line facilities	Bell Atlantic			
PFSAS	Private line facilities	Bell Atlantic			
PFSBL	Private line facilities	Bell Atlantic			
PFSFL	Private line facilities	Bell Atlantic			
PFSFS	Private line facilities	Bell Atlantic			
PK4	PBX main station line off premises		BellSouth		
PLP	Centrex systems (including dormitory) having			Southwestern Bell	
PMWD2	Basic 2-wire data circuit	Bell Atlantic			
PMWD4	Basic 4-wire data circuit	Bell Atlantic			
PMWGX	Type c-off premises extension	Bell Atlantic			
PMWV2	Off premise extension	Bell Atlantic			
PMWV4	Voice circuit - basic 4 wire	Bell Atlantic			
PTBVL	Voice grade	Bell Atlantic			
PTK2X	PBX trunk enhancements-premium service 2 wire		BellSouth		
PXBZZ	Extension/s	Bell Atlantic			

USOC	Explanation	LEC which utilizes the USOC			
PTCHS	Intralata private line service-voice grade, interexchange 2231		BellSouth		
PYCMS	Intralata private line service-voice grade, interexchange 2463		BellSouth		
PYCQS	Intralata private line service-voice grade, interexchange		BellSouth		
PYKAL	Intralata private line service-sub voice grade, intraexchange		BellSouth		
PYKHL	Intralata private line service-sub voice grade, intraexchange		BellSouth		
RCFVE	Remote call forwarding	Bell Atlantic	BellSouth		
RCFVQ	Remote call forwarding		BellSouth		
RCFVS	Remote call forwarding	Bell Atlantic			
RCRC1	Miles(unbillable)			Southwestern Bell	
REFBN	Service connection basic number referral service				Pacific Bell
RGE	Feature-per station (restructured)			Southwestern Bell	
RJ11C	Miniature jack	Bell Atlantic	BellSouth	Southwestern Bell	
RJ14C	Jack-2 line modular baseboard type		BellSouth	Southwestern Bell	Pacific Bell
RJ1DC	Miniature jack	Bell Atlantic			
RJ21X	25-line connector jack	Bell Atlantic	BellSouth	Southwestern Bell	
RJ2GX	8-tie line jack	Bell Atlantic	BellSouth		
RJ31X	Series jack	Bell Atlantic			
RJ36X	Series jack	Bell Atlantic	BellSouth		
RJ41S	Single-line jack	Bell Atlantic	BellSouth		

USOC	Explanation	LEC which utilizes the USOC			
RJ45S	Jack-data, programmed single line interface to modem		BellSouth		
RJ48C	Service connection jacks(s) excluding labor chgs				Pacific Bell
RJ48S	Jack-data or digital 2 line, 8 position		BellSouth		
RKY	Plexar express-station			Southwestern Bell	
RTVX7	Blocking service charge	Bell Atlantic			
RTVXA	Blocking service charge	Bell Atlantic			
RTVXP	Option	Bell Atlantic			
RX2	Centrex line	Bell Atlantic			
RXRA1	Rate change primary station line .92 per month				Pacific Bell
RXRA2	Rate change primary station line .92 per month				Pacific Bell
RXRA3	Rate change primary station line .92 per month				Pacific Bell
RXR	Centrex line	Bell Atlantic			
S1BF1	Voice mail mailbox				Pacific Bell
S5DBD	Dual tone multifrequency pulsing option on DID		BellSouth		
SAU	Intralata private line service-type B signaling arrangement		BellSouth		
SAY	Intralata private line service-type C signaling arrangement				
SBS	Summary billing establishment	Bell Atlantic			
SD6	Switchway service	Bell Atlantic			
SDH1A	Switchway-service	Bell Atlantic			
SDNA1	Rate change centrex is primary station line .53 per month				Pacific Bell

USOC	Explanation	LEC which utilizes the USOC			
SDS	ISDN basic exchange service	Bell Atlantic			
SEQ1X	Wire maintenance plan	Bell Atlantic	BellSouth		
SESAX	Service establishment plan	Bell Atlantic			
SESCL	Service establishment plan	Bell Atlantic			
SLM	Intralata private line service, E&M signaling arrangement		BellSouth		
SOF	Service-charge to establish mail box	Bell Atlantic			
SRG	Selective class of call screening	Bell Atlantic	BellSouth		
SXS	Intellipath digital	Bell Atlantic			
TB2	Direct incoming trunk	Bell Atlantic			
TBACX	Combined trunk	Bell Atlantic			
TBAOX	Direct dial trunk	Bell Atlantic			
TCG	Trunk	Bell Atlantic			
TCM	Outward service trunk	Bell Atlantic			
TCPCX	Centrex iii/plexar-access line-flat combination			Southwestern Bell	
TCPCX	Centrex iii/plexar-access line-flat combination			Southwestern Bell	
TDD1X	PBX service, DID inward trunk		BellSouth		
TDN	Touch-tone	Bell Atlantic			
TDT1X	Direct incoming trunk	Bell Atlantic			
TDYCX	Trunk	Bell Atlantic			
TDYOX	Outward service trunk	Bell Atlantic			
TEW25	Fire retardant cable	Bell Atlantic			
TEWO2	Fire retardant cable	Bell Atlantic			
TEWO4	Fire retardant cable	Bell Atlantic			
TF5OX	Leaky PBX-outward trunk, flat rate measured usage		BellSouth		
TFBO2	Trunk-direct inward dialing to stations-flat rate			Southwestern Bell	

USOC	Explanation	LEC which utilizes the USOC			
TFBO2	Wink start 4 digit acc dtms			Southwestern Bell	
TFBO3	Local exchange usage-trunk-flat-two way			Southwestern Bell	
TFBCX	PBX service combination trunk		BellSouth		
TFC	PBX service, combination flat rate trunk		BellSouth		
TFN	PBX service, inward flat rate trunk		BellSouth		
TFU	Local exchange usage - trunk-flat-out dial-per			Southwestern Bell	
TJB	Touch-tone trunk	Bell Atlantic	BellSouth	Southwestern Bell	
TMBCT	Message service	Bell Atlantic			
TMB1T	Message service	Bell Atlantic			
TSW	Charge-smulated access line where PBX trunk			Southwestern Bell	
TSW	Charge-smulated access line where PBX trunk			Southwestern Bell	
TK91X	PBX business service inward only trunk local optional service		BellSouth		
TK9CX	PBX business service combination trunk, local optional service		BellSouth		
TTB	Touch-tone business	Bell Atlantic	BellSouth	Southwestern Bell	
TTM1X	Area calling service-business trunk lines-inward only-economy		BellSouth		
TTPCX	Area calling service-business trunk lines-both ways-standard		BellSouth		
TTPOX	Area calling service-business trunk lines-outward only-standard		BellSouth		

USOC	Explanation	LEC which utilizes the USOC			
TTR	Touch-tone residence		BellSouth		
TW6	Direct incoming trunk	Bell Atlantic			
TWD1X	Direct incoming trunk	Bell Atlantic			
TXG	Combined trunk	Bell Atlantic			
TXM	Direct dial trunk	Bell Atlantic			
UA7HW	Customer or vendor owned complex wire			Southwestern Bell	
UF7	Sliding scale charge	Bell Atlantic			
UNR	Numbering/automat	Bell Atlantic			
UPPE1	Area calling service-usage package economy		BellSouth		
UPPO1	Usage package, economy service option, per line		BellSouth		
UPPO2	Usage package, standard service option, per line		BellSouth		
UPPS2	Area calling service usage package-standard		BellSouth		
URS	Other fee payment			Southwestern Bell	
URT	Unrestricted-standard feature-perstation			Southwestern Bell	
VF3	Foreign exchange service	Bell Atlantic			
VMN2X	Service-voice mail	Bell Atlantic			
VMS7O	Callnotes-general business			Southwestern Bell	
VOP3X	Large volume discount 23%	Bell Atlantic			
VOP5X	Large volume discount 25%	Bell Atlantic			
VSFAZ	Call answering	Bell Atlantic			
VUM24	Megalink/lightgate (aka BellSouth spa pt to pt network)		BellSouth		
W8B1T	Custom 8-business 15.00 per month				Pacific Bell
WMR	Access line-deregulated			Southwestern Bell	

USOC	Explanation	LEC which utilizes the USOC			
XFB	CPE-PBX-business flat-class of service			Southwestern Bell	
XLBXX	PBX service measured service, dial, customer owned equipment		BellSouth		
XP8XX	PBX business service economy service option, customer provided		BellSouth		
XU22X	Mileage-flat rate	Bell Atlantic			
XUW1X	Mileage charge	Bell Atlantic			
ZZYEB	Direct inward dialing toll	Bell Atlantic			

Telecommunications Expense Management

Appendix B

List of IXC PIC Codes

IC Name	PIC	
Allnet Communications Service	ALN	444
American Long Lines	ALG	241
American Network Exchange	ANK	370
ATC Long Distance-DDS	MIC	789
AT&T Communications	ATI	732
AT&T Communications	ATX	288
AT&T Communications	AWD	387
ATX Telecommunications Svcs.	ATZ	004
Aus Inc	AUU	301
Bell-Save	BLT	682
Blue Ridge Telephone	TPO	870
Business Telecom Inc.	BTM	833
Cable & Wireless Communications, Inc.	TDX	223
Cable & Wireless Communications, Inc.	CWI	839
Capital Network System, Inc.	CAQ	425
Capital Telecommunications Inc.	CPL	221
Chadwick Telephone	CWV	909
Cherry Communications, Inc.	CHY	270
Cleartel Communications	CRZ	584
Coast International, Inc.	CIZ	063
Comm. Telesystem's International	CXM	502

IC Name	PIC	
Commonwealth Long Distance Co.	CWZ	336
Conquest Operator Service	CQO	319
CPS Operator Services Inc.	CFP	309
Digitran Corporation	DTR	543
Eastern Telecom Corporation	EAS	136
Eastern Telephone Systems	ETS	054
Excel Telecommunications, Inc.	EXL	752
IDB/World Com	FTC	458
Innovative Communications, Inc.	INV	510
ISI Telecommunications	ANU	145
Keystone Long Distance, Inc.	KYL	699
LCI International/LITEL	LGT	432
LDDS Communications, Inc.	LDD	450
Lexitel	LEX	066
Long Distance Svc, Inc.	LDW	537
Long Distance Tel Savers, Inc.	LTS	213
Long Distance Wholesale Club	LWH	297
Matrix Telecom	MXT	780
MCI Telecommunications	MCI	222
MCI Telecommunications	MCJ	088
MCI Telecommunications	MCK	898
MCI Telecommunications	MCG	888
Metrocomm	TWH	860
Metromedia/Communication Corp.	ITT	488
Metromedia/Communications Corp.	ANW	311
Metromedia Long Distance	MTD	011
Mid Atlantic Telecom	MAD	066
National Telephone Exchange, Inc.	NLE	746

IC Name	PIC	
Nche Telecommunications Network, Inc.	NCE	017
Net Express Communications	NOE	135
Network One	NXC	388
Oncor Communications, Inc.	ITG	805
Oncor Communications, Inc.	ONR	658
Opticom	SVL	880
Pace	PAC	757
Phone America	BML	742
Polar Communications Corp.	PLR	967
Professional Communications, Inc.	PRM	726
RCI Corp	RTC	211
Shenandoah Long Distance Co.	SHL	207
Sprint Communications Company	UTC	333
Sprint Hospitality Group	LDU	252
Startec, Inc.	STZ	719
Target Telecom, Inc.	TAG	995
Telco Comm. GRP DBA Dial & Save	TDG	457
Telecom* USA	SNT	852
Telecom* USA	TDD	835
Teledial America, Inc.	TED	040
Tele-Fiber Net	TFB	008
Telemarketing Investments, Ltd.	TAM	007
Telenet Communications Corp.	GTS	759
Telescan, Inc.	TZC	731
Telestar Communications, Inc,	TPD	837
Televue Corp.	SDY	707
Total-Tel USA	TTU	061
TRI *Tel Communications	TIQ	874

IC Name	PIC	
US Wats	UWT	200
Value Added Communications	VAC	817
Vartec	VRT	811
West Coast Telecommunications Inc.	WCU	569
Westinghouse Electric Corp.	WIN	946
Wiltel, Inc.	WTL	555

Appendix C

**10 Steps that Vendors can take
to Improve Customer Service
and Billing Accuracy**

1. All Customer Service Records (CSRs) should be printed on clean, columnar paper. They should not be local prints of computer terminal screens.

2. Customers should be educated and encouraged to review and obtain their CSR. Currently, most of the LECs & IXCs provide a general breakdown of the monthly billing charges once a year. This breakdown does not provide enough information to the customer, and in fact gives the customer a false sense of security. Many customers do not realize the amount of information that is available to them on the CSR.

3. CSRs should have an audit trail of activity for each USOC billed on a customer CSR. Currently, each time a rate change occurs, the activity date is overlaid with the rate change date. This confuses both customers and many telephone company representatives, as they erroneously believe an activity (meaning physical activity) has occurred on a particular line or circuit on a particular date. Changes

for lines or circuits are often not challenged as it appears physical activity is taking place on these services.

4. USOC information should be disseminated to customers and consultants on request. LECs & IXCs should offer courses on how to read their bills and CSRs in an effort to make their billing procedures easier to understand.

5. Special service charges should be itemized on a separate and distinct bill. Most vendors intermix charges for special service circuits with charges for POTS lines. When charges are intermixed, bills are hard to verify. Many companies have separate voice and data departments. Separate billing by the vendor will help these companies assign verification and payment responsibilities to the proper department.

6. Make customers aware of the benefits of receiving and paying their bills via Electronic Data Interchange (EDI).

7. Customers should have on-line access to telephone company tariffs via the Internet.

8. The LECs & IXCs should be required to have the customer sign an authorization form before they bill wire maintenance charges. These charges mysteriously appear on many bills, and often cause confusion and resentment when customers discover them. Most LECs & IXCs cannot tell you how some of these charges got onto their bills.

9. Instruct Customer Service Representatives not to use early retirement and/or layoffs as excuses for delays. With most of the LECs

& IXCs reducing their work force, representatives often tell customers that they are short-handed. A customer with a problem does not want to hear that the company is shorthanded, they want results and solutions to their problems.

10. LECs & IXCs should start internal auditing units that randomly verify billing (free of charge to the customer). They should be as knowledgeable and motivated as independent auditing firms. These units should also market themselves to customers and offer to provide applications engineering services to interested firms. The goodwill generated and the feedback obtained from these audits will prove invaluable to increasing customer loyalty and billing accuracy.

Glossary

Access – within the context of CABS, LATAs and other local telephone company/long distance carrier interfaces, access refers to the delivery of transmissions through the local exchange network to a POP.

ACD or Automatic Call Distributor – a telephone call management system designed to evenly distribute incoming calls to group service representatives and/or service support personnel.

Advanced Billing – Advanced billing pertains to the fixed cost portion of the bill, e.g. cost of non usage-sensitive services – line charges, touchtone, directory advertising, etc. With advanced billing you are paying the monthly charge for service for the upcoming month "in advance".

Arrears Billing – is billing for service by the telco for the previous month's fixed and/or usage-sensitive service. New England Telephone used Arrears Billing.

ATM - is an abbreviation for Asynchronous Transfer Mode. ATM is a high speed, high bandwidth, low delay, packet-like transmission technology.

Bandwidth – is the width or capacity of a communications channel.

BELLCORE – is an acronym for Bell Communications Research, a research company that was jointly owned by the seven RBOCs. BELLCORE set standards, coordinated USOC implementation and coordinated network services among RBOCs and Independent telcos. BELLCORE was subsequently sold and is now known as Telcordia Technologies. Its influential role among the remaining RBOCs is greatly diminished.

Billing Systems – are a group of telephone company computer systems responsible for creating your customer service record (CSR), and rendering your monthly telephone bill.

BOC – Bell Operating Company. There were twenty-two Bell Operating Companies that were folded into seven RBOCs created by the 1984 Bell System Divestiture Agreement.

CABS – Carrier Access Billing. A billing system used by the local telephone company to bill long distance carriers for access services.

Casual Calling – refers to the ability to use AT&T, MCI or Sprint's long distance network without designating one of them as your primary carrier. For example, to use AT&T you would prefix the number you want to reach with 10288. To use MCI's network, your would prefix the call with 10222, and to use Sprint's network, 10333. Casual Calling is not discounted.

CDR – Call Record Detail. CDR is a formatted report generated by a call accounting system that details call characteristics, e.g., time of day, length of call, carrier, called number, etc. CDR is sometimes referred to as SMDR for station management detail record.

CO or Central Office – A telephone switching center that provides dial tone, and routes calls through the public switched network. The first three digits of your telephone number are associated with the particular local central office that serves your area.

Centrex – Centrex is a business telephone switching service controlled through a local central office. It is basically a single telephone line service to individual phones (the same as you get at your house) with many built in features, including intercom, call forwarding, call transfer, toll restrict, least cost routing and call hold (available on single line phones). Centrex normally comes in two varieties; analog and digital.

CSU – a Channel Service Unit is CPE used to terminate digital-type service such as T1s.

CLECs – Competitive Local Exchange Carriers. A new local exchange company competing with an incumbent local exchange carrier or ILEC.

CO lines – are telephone lines that connect your CPE to your local telephone company's central office. They also connect you to the nationwide telephone-switching network.

Collocation – allows certain telephone company customers to rent space within the telephone company's central offices, and to use this space to set up and maintain their own facilities and equipment. This equipment can then be directly connected to the telephone company's network.

Collect Call – A telephone call generally requiring operator assistance. An automated or live operator asks the called party if they agree to be billed for the call.

Corridor Optional Calling Plan – is offered in certain regions in the United States. In New York, New York Telephone offers a discounted way for subscribers in the New York City area codes to call the five northern New Jersey counties – Bergen, Essex, Hudson Passaic and Union for a monthly charge. Using the special access code 10698 (10NYT) calls originating from New York City and going to these five New Jersey counties can bypass normal long distance carrier networks, e.g. AT&T, MCI, etc. The Corridor Optional Calling Plan is also available from the five New Jersey communities to the New York City. New Jersey Bell also bills these calls at a discount. The telephone number dialed would then be preceded by 10658 (10NJB).

CPE – Customer Provided Equipment.

CRIS – Customer Record Information System. This is the mainframe system that the telephone representative references to retrieve and review the Customer Service Record. CRIS is the system used by RBOCs to bill residence and business customers.

CSR – Customer Service Record. A CSR is a record of all your telephone company services and service order activity.

DDD – see MTS

Demarcation Point – The "demarc" is the point where the telephone company responsibility ends within a customer's premise. This termination point is normally a 42A connecting block or equivalent. The telco maintains the facilities between the demarc and the central office.

DSU or Digital Service Unit – a DSU is CPE that, along with Channel Service Units, are necessary for connecting digital T1s to the telephone company's network.

DID or Direct Inward Dialing – You can directly dial a particular extension within a company by using DID trunks, thereby bypassing the PBX attendant. Each extension is assigned a discrete seven digit number. DIDs allow only inward calls. You cannot dial out on a DID trunk.

Divestiture – On January 8, 1982 AT&T signed a consent degree with the Justice Department, stipulating that on midnight December 31, 1983, AT&T would divest itself of its 22 Bell Operating Companies. According to the terms of the agreement, the 22 Bell Operating Companies would be formed into seven regional bell operating companies of roughly equal size. The terms of the agreement placed various restrictions on AT&T and these new RBOCs. Specifically, the RBOCs were not allowed to provide long distance service, could not manufacture telephone equipment or provide information services. AT&T was not allowed to provide local telephone service in competition with the RBOCs. (Several restrictions have since been lifted).

DMS – Digital Multiplex System, a family of central office switches made by Nortel (formally Northern Telecom).

DOD – Direct Outward Dial Trunk

DS0, DS1 – Pronounced "D-S Zero" and D-S One". These are units of transmission as defined by bandwidth. DS-1 is T-1, at 1.544Mbps. A single DS0 is 64 Kbps. Twenty-four DS0s (24 x 64Kbps) equals one DS1.

E-zine – magazines published on the Internet

Econopath Calling Plans – Bell Atlantic NY's version of an intraLATA economy calling plan available for businesses to make calls within regional calling area (LATA). For example in New York, Econopath offers Manhattan businesses a discount for calls made to the East Suffolk (Long Island) region.

ESS – Electronic Switching System, a family of central office switches made by AT&T, e.g. ESS 1 and ESS 5.

Exchange Service – refers to a basic 1MB (Measured Business Line) or 1MR (Measured Residence Line) line connected to the telephone central office. This line provides dial tone and connects to the public switch network.

Facilities-Based Carriers – Telephone companies that have their own physical networks, as opposed to companies that resell services provided by other carriers.

FACS – Facility Assignment and Control System. This system is used to provision customer line orders.

FCC or Federal Communications Commission – The FCC regulates all interstate telephone traffic and international traffic originating within the USA.

Fiber Optics – thin strands of very pure glass through which light is transmitted. Fiber Optics offers high bandwidth, economies of scale and require minimal distribution space.

FID or Field Identifier - Used by RBOCs to further describe or define USOCs.

Flexpath – New York Telephone service that provides 1.544 Mbps service between a digital PBX and the servicing central office. Flexpath has Direct Inward Dial Calling (DID) and Direct Outward Dialing (DOD) capabilities.

Gold Number – a/k/a vanity number. It is a telephone number that is easy to remember, e.g. 767-1111. You pay an additional monthly charge for such a number.

ILEC – Incumbent Local Exchange Carrier. A term synonymous with the traditional local telephone company.

INOF – interoffice mileage charges. The mileage-sensitive charges between COs serving a particular circuit.

Intellipath Digital Centrex Service – Bell Atlantic-NY version of digital Centrex service. This is Centrex that is provisioned out of a digital CO.

IP telephony – Internet Provider Telephone uses the Internet to transmit telephone calls.

IPV – Invoice Processing and Validation. A software application that manages the processing and validation of telecommunications billing.

ISG – Information Strategies Group. Telecommunications billing experts providing IPV, consulting, and auditing services to telecommunications and financial managers.

IT or Information Technology - IT is a more current name for Data Processing.

IXC – an abbreviation for interexchange carrier.

LATA or Local Access and Transport Area – one of 161 local access service areas in the US. LATAs were formed as a result of the Bell System divestiture agreement to define the areas within which local telephone companies would operate. Long distance carriers would then be consigned to handle interLATA traffic.

Leased Line – A telephone line directly connecting two points. You rent a leased line typically by the month and normally do not pay extra for usage. It's yours to exclusively use. The terms Lease Line, Private Line and Tie Line are frequently used interchangeably.

LEC – A Local Exchange Carrier that is normally the traditional local telephone company in any given area. The terms LEC, telco ILEC and telephone company are often used interchangeably.

LMOS – Loop Maintenance Operation System. A computer system used by the operations departments of local telephone companies to ensure the proper performance of outside plant.

M & A – Mergers and Acquisitions.

MDF – Main Distribution Frame. A primary piece of switching hardware found in a CO.

MFJ or Modified Final Judgment. Within the context of the Bell System divestiture, the MFJ was essentially the Divestiture Agreement between the Department of Justice and AT&T.

Month-To-Month Billing – The standard way of paying for telephone service. Some telephone company services now come in "rate stability" packages. This means if you commit to the service for three to five years, you pay less each month.

MTS – Message Telecommunications Service. This is a regular telephone toll service, as distinguished from WATS service. It is "full fare" long distance service that is dialed on a 1+ rather than a 0+ basis. It is also known as DDD or direct distance dialing.

One Way Trunks – These trunks are ordered from the telephone company and can be used to receive incoming calls from the central office or for outgoing service to the central office, but not for two-way calling.

PBX or Private Branch Exchange – A private (you own it), branch (meaning it is a CO replica), exchange (switch). A PBX or smaller version called a key telephone system allows you to connect extensions to the PBX and allows for dialing from extension (station) to extension (station) without the need to involve the telephone company's central office. The PBX provides dial tone, talking battery and ringing generator in the same way the telephone company's central office does for phones directly connected to it. The PBX is in turn connected to the central office by way of central office (CO) trunks (lines) that allow an extension to make and receive calls outside of the PBX.

Private Line – A direct line or channel between two or more locations that are exclusively available to a particular customer 24 hours a day. This type of line does not have access to the switched network. Known also as a Leased Line and Tie Line.

POP – Point of Presence. A long distance carrier's connecting point with a local telephone company.

POTS – Another name for Plain Old Telephone Service. Just dial tone, no fancy features.

Provisioning Systems – This is a group of telephone company systems that are tightly coupled in that they process the service order request that is originated by the Service Order Processor (SOP) system. The provisioning systems assign cable and pair facilities as well as direct central office and field personnel to perform the physical work to satisfy the work order request. After the work is completed, the service order is so noted and the order is then sent to the billing system.

PTT or Postal Telephone and Telegraph – a term used to describe telephone companies outside North America.

Public Switched Network – refers to the network generally available to the public as opposed to PBX or other privately controlled networks.

RSP or Rate Stability Plan – a discount in exchange for a 3 to 5 year commitment term for local services. The customer is normally insulated from any rate increases occurring during the course of such a plan. It's a good plan to consider, but beware of penalties if you move during the term period.

RCF or Remote Call Forwarding – Telephone company service that instantly forwards a call to another prearranged number. RCF is a cheap way to establish a "presence" in a distant city.

Run-Rate – another term for monthly billing rate.

Service Order – A work order request issued by a telephone company's representative that details the type and cost of telephone service requested by a particular customer. Your service request is translated into USOC format and is entered into the telephone company's Service Order System (SOP) for provisioning and billing.

SMDR or Station Management Detail Record. See CDR

SMDR Port – Modern PBXs and some key systems have an Station Message Detail Recording (SMDR) electrical plug, usually a RS-232, 25-pin connector. The SMDR port is used by call accounting systems and some toll fraud detection units. The telephone system sends information (normally referred to as Call Detail Records) on each call that is made and received. This record normally contains the extension that originated the call, the trunk that handled the call, duration of the call, time and date stamp and pricing information.

SOP – Service Order Processing. The function of creating and processing customer requests for telephone services.

Summary Billing – Telephone company service that itemizes the charges for all your different locations onto one bill.

SUPERPATH – New York Telephone's point-to-point T1 service.

Switchway – New York Telephone's switched service that lets you dial up and transmit data at 56Kbps. It is end-to-end digital service.

T-1 – Also spelled T1. This is a digital transmission line with a capacity of 1.544 Mbps (1,544,000 bits per second). A T1 uses two pairs of twisted wires, the same kind of wire you would find in your home. A T1 can normally handle 24 simultaneous voice conversations with each digitized at 64Kbps. With new voice encoding techniques, it can sometimes handle even more voice conversations. T1 links can be connected directly to new digital PBXs.

Telephone Exchange – A telephone company switching center for connecting and switching phone lines. A European term for what North Americans call a central office.

Third Party Call – Any call charged to a number other than that of the origination or destination party.

Tie Line – A dedicated circuit that connects two PBXs.

Tip & Ring – An old fashioned way to refer to a basic telephone line with dial tone, but the term is still used throughout the telephone industry. Ring is normally the red wire and the tip is green wire with ground being yellow wire. Tip also refers to the transmit side and Ring to the receive side of a circuit. The genesis of this term stems from the old operator cord boards, which had plugs with "tips" and "rings" representing each contact point.

TIRKS – Trunk Integrated Record Keeping System. A mainframe system that assigns and tracks facilities.

Tone Dialing – A push button phone that emits a different sound (frequency) for each digit or special function appearing on the touch tone key pad.

Touch-tone – A trademark owned by AT&T for tone dialing.

Trunk – A communications line between two switching systems. A central office (CO) trunk connects the customer's PBX switch to the telephone company local serving central office. A tie trunk (or line) connects two or more PBXs together for passing voice and data between them.

Twisted Pair – Two insulated copper wires twisted around each other over the length of the cable to reduce induction noise. Twisted pair is also synonymous with local loop or subscriber loop used to connect the local subscriber's equipment (e.g. PBX, telephone set, etc.) to the local serving central office.

Two Way Trunk – A trunk that is used for two-way conversation into or out of a telephone system. Some trunks are used as one-way for either receiving calls or making calls but not both.

Usage Based – Refers to a rate or price for telephone service based on usage rather than a fixed monthly rate. An exchange line is charged a usage rate based on the number of calls made and also for the duration of each call.

User Loop – A 2 or 4 wire circuit that connects a user to the central office.

USOC – Universal Service Order Code. (Pronounced "U-Sock"). A Bell System coding protocol identifying particular services or equipment. The USOC code is a remnant from the old Bell System days where AT&T set standards for all of the 22 BOCs. After divestiture, BELLCORE assumed responsibility for setting standards for the RBOCs. The idea behind using USOCs is to identify every service or product offered by a telephone company with a discrete code that can be read and interpreted by sophisticated service order and billing systems. There are over 30,000 USOCs in use by the RBOCs and other telcos.

V & H – Vertical and Horizontal grid coordinates. V&H coordinates are put through a mathematical equation to determine "airline distance" between rate centers (central offices). Each central office is identified by its own unique V&H coordinates. They are used to calculate mileage billing on leased lines and other mileage sensitive services.

Voice Circuit – A typical analog telephone channel coming into your house or office. It has a bandwidth between 300 Hertz and 3000 Hertz, which is sufficient to audibly recognize and understand human voice.

VPN – Virtual Private Network. A service offered by the long distance carriers to large companies that wish to link multiple locations without having dedicated physical lines between those locations.

WATS – Wide Area Telecommunications Service. Basically discounted toll service provided by all long distance and local telephone companies. AT&T started WATS but forgot to trademark the name, so now every supplier can use it. There are two types of WATS services; in and out WATS, i.e. those WATS lines that allow you to dial out and those on which you receive incoming calls.

X.25 – a defined communications protocol used mainly for packet switching.

Yellow Pages – A directory of telephone numbers classified by the type of business. It was printed on yellow paper throughout the 20th century. Yellow Pages is not a registered term and is used today generically.